THE BACK HOME SERIES

Series Titles

Wrong Tree: Adventures in Wildlife Biology
Jeff Wilson

At the Lake
Jim Landwehr

Body Talk
Takwa Gordon

The In-Between State
Martha Lundin

North Freedom
Carolyn Dallmann

Ohio Apertures
Robert Miltner

Praise for

Wrong Tree

"Jeff's wild and often amusing lifestyle, captured in this book, shows that he is the last of a breed. But our movement, environmentalism, is not dying and Jeff offers a beacon for the next generation to preserve its ties to the land."

—Rocky Barker
Pulitzer Prize Finalist
author of *Scorched Earth: How the Fires of Yellowstone Changed America*

"As a DNR Wildlife Technician, Jeff Wilson spent 21 years banding ospreys, helping manage beavers, dealing with bears and so much more. His love for animals began during his growing-up years on an Iowa farm when he had wild animals as pets. Especially memorable was his pet fox, Zorro. Here is a well-written story of love for the outdoors and especially its wildlife."

—Jerry Apps
author of *Whispers and Shadows*

"Jeff Wilson…is a world class individual and has more stories than anyone I have ever met."

—Jim Paruk
author of *Loon Lessons*

"I wouldn't have believed half of the stories in this book if I hadn't been there myself!"

—John Olson
Retired Furbearer Specialist and Wildlife Biologist
Wisconsin Department of Natural Resources

"I always knew Jeff Wilson could tell a good story, but this book proves he can write one too."

—Jeff Richter
photographer, *White Deer*
owner of Nature's Press

"I was a fellow wildlife professional with Jeff Wilson and many of the stories touched very close to home for me. However, anyone who enjoys wildlife and the great outdoors will be captivated by the stories in *Wrong Tree*. Jeff's adventures are entertaining and educational. The experiences of a real life field biologist are cleverly blended with ecological knowledge and important events in Wisconsin wildlife conservation to create wonderful stories. Terry Daulton's artwork is a valuable bonus."

—Scott Craven
Professor emeritus, Wildlife Ecology
University of Wisconsin-Madison
Member, Wisconsin Conservation Hall of Fame

"I loved reading *Wrong Tree*, but I also loved the illustrations. Great art is a perfect compliment to great stories—kudos to Terry Daulton for her fine eye!"

—John Bates
Naturalist and Author
Manitowish River Press

"Jeff's book shows passionate concern for the wildlife of the Midwest, North America, and the future."

—Ellen Akins
author of *Home Movie*, *Little Woman*, and *Hometown Brew*

"Jeff's book is an illuminating and entertaining journey of inquiry into the wonders of nature."

—Tia Nelson
environmentalist
daughter of Earth Day founder Gaylord Nelson

Wrong Tree

Adventures in Wildlife Biology

By Jeff Wilson
Illustrations By Terry Daulton

Cornerstone Press
Stevens Point, Wisconsin

Cornerstone Press, Stevens Point, Wisconsin 54481
Copyright © 2022 Jeff Wilson
Illustrations © 2022 Terry Daulton
www.uwsp.edu/cornerstone

Printed in the United States of America by
Point Print and Design Studio, Stevens Point, Wisconsin

Library of Congress Control Number: 2022945452
ISBN: 979-8-9861447-8-8

Book design by Jerry Lehman

Cornerstone Press titles are produced in courses and internships offered by the Department of English at the University of Wisconsin–Stevens Point.

DIRECTOR & PUBLISHER   EXECUTIVE EDITOR
Dr. Ross K. Tangedal   Jeff Snowbarger

SENIOR EDITORS
Lexie Neeley, Monica Swinick, Kala Buttke

PRESS STAFF
Patrick Fogarty, Brett Hill, Amanda Leibham, Abbi Wasielewski

To my mentors,
who taught me everything from snaring bears to dynamiting
beaver dams, rapelling from treetop nests or examining deer
molars to determine age.

This dedication is in thanks for all those lessons afield.

Table of Contents

Preface   xiii

Prologue   xv

Creek Roaming Days   1

Nature's Engineers, Friend or Foe?   11

Darwin Awards   33

The Scent of Skunks   45

The Elephant in the Room: Human Dimensions   53

(Lions and Tigers) and Bears Oh My!   87

Superior Bears   103

The Problem with Bears   119

Osprey Penthouse Apartments   147

Loon Research Days   169

It's in Their Blood   189

High-Tech Loons   201

Economic Decisions, Claire, and the Butterfly Effect   223

Innocents Abroad   243

A Corn Farmer's Funeral   259

Epilogue – Keep the Fire Burning: A Visit with Leopold's Ghost   263

Acknowledgments   271

Partial Cast of Characters   273

Preface

"Said one of these men long past 70 years of age 'I could carry, paddle, walk and sing with any man I ever saw. I have been 24 years a canoe man and 41 years in service, no portage was ever too long for me, 50 songs could I sing. I have saved the lives of ten voyageurs; have had 12 wives and six running dogs. I spent all my money in pleasure. Were I young again, I would spend my life the same way over. There is no life so happy as the voyageurs' life."

—Grace Lee Nute, *The Voyageur*

Though I've always felt a link to this hard-working Frenchman, along with a secret wish that I had been alive to see those early years of the fur trade in North America, I know that he and I aren't really alike. For one thing I've had far fewer wives. Also, I didn't spend all my money on pleasure, just a good bit of it. I can however say that no portage has yet been too long for me, and my combined time working with various resource agencies throughout the U.S. and Canada put me in service about 32 years. As my family will attest I definitely lack in the singing department, so in this book I intend to replace the songs of my life with stories. Here I recount my adventures with wildlife, from the Iowa prairies of childhood to Wisconsin's northwoods and beyond. They are my voyage—and by the way, were I young again, I would spend my life the same way over—there is no life as happy as the voyageur's life.

Prologue

I was 40 feet up a massive white pine, my fingers gripping the rough bark. My feet felt secure, balanced on the metal tree climbing spikes that buckled over my boots and around my calves. Just above my head were the first branches that would allow me handholds, so I released one side of my belt harness and let it hang as I reached to grab the lowest branch and heaved myself up and into a sitting position on the limb. Glancing down, far below, I could see the 12 field-trippers and my eagle-banding partner, Bruce, peering up intently. Just past the crew, our boat stern peaked out from the willow shrubs lining the shore of Blair Lake. I could hear Bruce explaining the banding protocol, entertaining the group who had signed up for this tour with the Natural Resources Foundation. They expected a good show and we would deliver.

The group had been a rapt audience from the start. We'd gathered that morning at Springstead Landing, a boat launch on the Turtle Flambeau Flowage in northern Wisconsin. They'd watched the shores with interest during our circuitous boat ride, skirting islands and wetland channels. When I unpacked the climbing gear—demonstrating the harness around my waist and legs, and how the flip belt would secure my progress up the trunk—they'd asked good questions. I even passed around my climbing spikes so they could feel the sharp point that would stab into the bark and enable my ascent up the pine.

Clearly they found our risky occupation novel, if not a bit crazy, but to me, this tree was like an old friend. I had grappled up to its 80-foot crown before, and now with the branchless section of the trunk behind me I could relax a bit and enjoy the ascent. The nest was about 10 feet below the treetop in a thick fork. I climbed methodically until I was just below the nest, then strapped in with my flip belt again. This would allow me to maneuver safely around the bulky nest. It was a huge structure, four-foot wide and six-foot long. In particular, I needed both hands free in case there was a "jumper," a chick with big ideas who decided to test its wings a bit too early.

I repositioned my spikes, edged up and peeked over the rim of the nest. One chick hunched down, flat and motionless, but the other stood up, spread its wings and hissed before retreating to the far corner of the nest bowl. I began a one-way conversation with them, murmuring in what I hoped was a soothing manner as I slowly stretched up and halfway into the nest. I smiled to myself gauging that the chicks were the perfect age for handling and banding, about six or seven weeks. Most of their downy plumage was gone and their primaries were half "blood quill" and half feather. This meant that the Foundation crew could handle them safely, the chicks being robust enough for the process, and the humans relatively safe from beaks and talons. The chicks' feet were anatomically able, with full talons, but they did not yet know how to use them defensively.

I unclipped a zippered canvas chick bag from the carabineer on my climbing belt and reached for the hunched down chick first. I folded its wings tight against its body, and then inserted it headfirst, turning its body upright after the tail feathers cleared the mouth of the bag. I zipped the bag carefully, clipped it onto a rope and carefully lowered the chick down through the branches. The second chick took more finesse. I uncurled a roll of stout wire from my fanny pack, bent one end into a hook, slowly extended the wire across the nest, and slid the hook around the chick's legs, gently pulling it into reach. I grabbed it, safely folding its wings and repeating the bagging and lowering process.

Bruce shouted up, "All clear!" I would have 20 minutes or so before the ground crew would be done measuring, taking feather samples, and banding the chicks so I pulled myself up into the nest. I looked out at the view the eagles enjoyed every day and wondered how many humans have had such an opportunity. The floor of the nest where I sat was lined with soft grasses and a few white pine sprigs. It was clean, the chicks always back up to the edge of the nest to defecate, sending a stream of whitewash out and down to the forest floor. But there certainly was an aroma, fishy and sharp, and part of a northern pike was draped over the sticks on the far side of the nest. I took a few moments to dig in the nest duff, looking to see if there were unhatched eggs, as I knew adult eagles often buried infertile ones.

My brief survey complete, I turned my gaze farther afield. I could see water in every direction for miles, dotted with islands, wetlands, and stump fields. I could make out forest composition as well. Glimmers of white peaked from paper birch stands, young aspen glowed a brighter green, and the crowns of old growth hemlock and pines stood dark green against the sky. I could even see a loon foraging on the lake below.

I thought back to the first time I had seen an expansive view like this. I was only 14 years old and I had climbed up the rickety ladder on the side of our silo at the family farm in Iowa. My brother and I wanted to raise a pigeon as a pet, and we had spotted a nest near the silo's rounded top. Climbing the silo was not something our parents would have approved, so our trembling legs were probably part fear at the exposed rusty ladder rungs, part fear of falling, and the certain knowledge that punishment might result from our escapade. But when we got to the top and carefully tucked one of the squabs into a jacket pocket we hung there for a few minutes taking in the panorama—fields, hedgerows, a far off tractor working the land—topped by that expansive Iowa sky.

A call from below brought me back. I had travelled a long journey from that Iowa silo to this eagle nest in the northwoods of Wisconsin and it was time to get back to work. I started carefully pulling the chicks back up to their lofty home. There would be time for recollections when my feet were back on the ground.

Creek Roaming Days

"There are some who can live without wild things and some who cannot."

—Aldo Leopold

We were eating lunch when the door flew open and the Fuller Brush salesman rushed inside yelling, "Get a gun…it's in the yard…a rabid fox and he almost got me." I looked out the window and sure enough there was a red fox standing next to our tractor. I glanced at Mom and, before running outside said, "Don't worry. I know what to do." Mom and the salesman watched out the kitchen window. As soon as I left the porch the fox raced after me in hot pursuit. We circled under the trees and around shrubs until the fox caught up to me. I stumbled and fell in the grass and the fox jumped on top of me. We tussled for a few moments, then I picked up Zorro, my pet fox, and carried him to the porch to show the salesman. My mother tried to chide me for terrifying the man, but she had a twinkle in her eye and soon we were all laughing. This wasn't the first time I had performed this stunt.

In retrospect, my childhood seems storybook lucky. In the Iowa countryside of my youth, farming was the way of life. Every kitchen seemed to have a fresh-baked pie, and a kid's "home range" was measured in miles, not blocks. My brother Brad was two years younger and my sister Janice was three years older. We loved working on the farm and started along-side our parents as soon as we were strong enough to lift a hay bale and had legs long enough to reach the clutch pedal of the Farmall H tractor. I was small for my age, which meant that I had to learn to drive the tractor starting in second gear instead of first. My legs were too short to push down the clutch when the gearshift lever was way over to the left side for first gear. Because I was the oldest boy, as soon as I was big enough and old enough to start operating farm machinery, I took over Janice's fieldwork, freeing her up to help Mom in the house and garden. We still all worked together on big projects like planting and harvesting the crops, baling hay, castrating calves, docking lamb tails, and our never-ending summer job of walking the soybean rows with machetes, cutting out volunteer corn that had re-sprouted from last year's crop and pulling buttonweeds. These last jobs are now done by herbicides.

Brad was the athlete and musician among us. He was on the basketball, football, and track teams and played guitar and sang and wrote music. Lacking Brad's talents and physical attributes, I found my calling in roaming the rural landscape, hunting, fishing, trapping, and exploring the creeks, rivers and marshes around our farm in eastern Iowa—adventures that inspired one of Brad's songs, "Creek Roaming Days." Our farm near Springville had a big red dairy barn and a silo surrounded by livestock pens, grain bins, a hog house, an apple orchard, and 340 acres of corn, oats and soybeans, as well as hay and pasture.

Both my parents came from farm backgrounds and would tell stories of how my grandparents farmed with horses, and how they lived without electricity, canning and storing their food in root cellars and pumping their water with a windmill. These stories inspired another of Brad's songs I love, "Two Pigs on a Running Board." He didn't sing it often because every time he did my mother teared up. The song's lyrics were about her dad, Grandpa Frank Marshall, losing their farm on two different occasions during the Great Depression after the bank foreclosed. My mom remembers the family loading the two pigs, all they had left, into wooden crates, strapping them to the running boards on each side of their Model A Ford, and driving away to try again, this time as farm renters instead of owners.

My dad's story was a little different. His family farm was about 50 miles away from Springville, near Dysart, Iowa. He was the youngest of seven kids (four boys and three girls). Two of his brothers went into farming. Dad enlisted in the army in 1942, was injured by a mortar blast in Germany, and ended up in a hospital in England. He recovered but carried shrapnel in his leg for the rest of his life.

After the war, his dad and brothers didn't have the money or land for Dad to join them, so he opted to attend college, majoring in business administration. After college he worked for a trucking company and eventually moved up to vice-president. He often took business clients to lunch at a local diner, Bishop's Buffet, and one day he met my mother, a bubbly brunette who waited on tables there. They began dating, and within two years they married and built a brand-new house in the suburbs of Cedar Rapids. By any 1950s social measure, they were headed for a comfortable middle-class life, but beneath the surface Dad and Mom were not happy. They were just not cut out for city life.

Grandpa Marshall told them about a farm for sale a mile down the road from his place outside Springville. It was eighty acres with an old house, a chicken coop, outhouse, and windmill. No other buildings, no barn, no machine shed, no silo. Grandpa would help with a down payment on the mortgage. Mom and Dad talked it over. It would mean selling and leaving Mom's new modern house and moving into a dilapidated old farm house and using an outside toilet. They decided to take the leap and soon they were farming corn, soybeans, and hay, with cows, chickens, and pigs on the side. Later they bought adjoining farms and Mom got a better house. They were living the good life working the farm together.

It was on the farm, during my "Creek Roaming Days," that I became fascinated with wild animals. Today it's illegal to keep any wild animal as a pet without a special permit but when I was a kid, it wasn't an issue. People's views on the treatment of wild animals were different then. We kids had all seen Walt Disney films featuring pet wildlife. Flipper the tame porpoise, and Gentle Ben—the bear that helped catch bad guys riding the front of Everglades Warden Wedloe's airboat—were popular television shows. I spent my life on the farm around young domestic animals, but wild animals were mystical and irresistible to me, and the idea having one for a pet simply filled me with joy.

Our first wild pet was given to us by Don Yanda, our neighbor up the road. He had a pet skunk named Flower whose scent gland had been removed. Don's two boys were grown and gone, so Brad and I spent a lot of time visiting Flower—she was cuddly and only had a faint odor. One day Don called Mom and offered to give us Flower. "They spend more time with her than I do," he said, so Flower became our pet. We eventually quit keeping her in a cage, giving her free run of the farmyard. She took her meals with the farm cats, and the dogs steered clear. If she got upset she would hiss and raise her tail as if to release her scent. It was a bit of a dud as she didn't have anything to squirt.

De-scented skunks were apparently a thing at the time. A man I met many years later—on a crew trapping elk in Wolverine, Michigan for their reintroduction into Wisconsin—claimed to have de-scented hundreds of skunks. It was a profitable business for him in the 1960s when de-scented skunks were sold nationally as pets. I recall seeing ads selling them in *Fur-Fish-Game* magazine, one of my favorite childhood reads. He wouldn't tell me how to perform the operation. The skunk pet trade crashed, he said, because word started to spread that skunks could carry rabies.

After skunks came the foxes. In the soil bank days in the 1960s the federal government paid farmers to set aside grassland acres, keeping them out of grain production. The only requirement was to mow the acreage once a year to keep the weeds down. This program was intended to help stop soil erosion and build better soil quality. It also created some incredible wildlife habitat supporting ring-necked pheasants, bob-white quail, cottontail rabbits, woodchucks, mice, ground squirrels and, of course, red fox. Every section of land had a fox den, or two, and our farm was no exception. The fox dens were often located on a sandy bank, rock pile, or brushy draw and usually were reconstructed from old badger or woodchuck burrows. They always had at least two exits and were easily distinguished by the tell-tale scattering of bones and feathers of various prey.

One March I was running the tractor on Grandpa's farm, discing last year's cornfield to level it for planting, when I spotted a fox den. I'd been discing for hours, riding the tractor and watching the sun sink lower. About dusk a vixen (female) fox came out of a nearby hedgerow and walked to the den. As I watched, eight fox kits came out and greeted her. I jumped off the tractor and ran to the den and was able to grab the last kit before it dove down the hole and escaped underground.

It was small as a kitten and I bottle-fed it for about a week before it would drink on its own. My uncle Stewart named it Zorro (he sold popcorn in Mexico and knew the Spanish word for fox was Zorro). We had just acquired a new puppy, and Zorro and the pup got along well. They ate and played together, which may be why Zorro became so tame. We didn't cage him, he had the run of the farmyard, and he learned to come when we called his name. The pup and the fox would curl up and sleep together. Zorro loved to ride in the truck with me, standing on the seat with his two feet on the dashboard looking out the window. If

I gave him some meat he would often grab it and run off a ways, then dig a hole and bury the food. He would carefully cover it with fresh dirt by using the side of his nose, and then urinate next to it. This habit of caching food is shared by other canids, like coyotes, wolves, and even dogs.

One day I was getting into the truck to drive away and I didn't see Zorro coming. As I was shutting the door of the truck Zorro leapt in to go for a ride but his head got caught in the door. We thought he was going to die. He was still breathing but unconscious so I laid him in a cage, gave him food and water and waited. In those days, farmers rarely took pets to the vet. We sure couldn't afford a vet bill, especially for a wild fox so it was not an option. He lay there for two days, and then started to recover. I nursed him and I often wonder if this experience bonded him to me, as he was as tame as a dog and I could pick him up and carry him.

We played together, especially a chase game where he would run off with a glove, cap, or anything I dropped and I'd have to chase him to retrieve it.

Zorro eventually started roaming farther and returning less often to be fed, until he returned to his wild nature completely. I felt okay with this, as I knew if he were to survive in the wild he had to develop a fear of humans. After Zorro I had two other pet foxes. One was pretty tame, but the other bit me, his canines going right through my hand, so I took him back and released him by his natal den. None of them ever came close to being as tame and friendly as Zorro.

But my creek roaming pets weren't confined to skunks and fox. I had read the book Rascal, by Sterling North, about a boy and his pet raccoon and I wanted a raccoon in the worst way. One day a neighboring farmer, Vernon Horak, said he had a raccoon den in an oak tree in his yard. After school I grabbed my buddy Kent Ulrich and we headed to the Horak farm to investigate. Kent volunteered to climb up the big oak and see if he could see anything in the dark cavity we spotted about 20 feet up the trunk. We used a ladder, as there were no low branches for him to hold, and when he peered into the hole he saw three kits. He pulled out two of them, handed them down to me and we each had a pet raccoon. His was a female and never became completely tame. It had a longer snout than mine and had a nasty habit of biting him all the time and nervously chewing on its own tail. That poor raccoon's tail eventually got infected and when it finally healed all that was left was a stub. My raccoon was a male and had completely different disposition and a handsome face. He looked friendly and kind and had a personality to match.

After bottle-feeding Cooney (I must've used up my talent for naming with Zorro) for a couple of weeks, we let him leave the cage and live outside. Mom had a rule in those days, no animals inside. Farm dogs had dog houses or slept in the barn. Cats lived in the barn. These were the days before dogs and cats were liberated and moved into people's houses, sleeping in their beds and on their furniture. It was before pet health insurance, pet crematoria, and pet cemeteries. If a dog or cat got old and debilitated a grave was dug, a shot was fired, and that pet was put out of its misery, so to speak.

Like Flower the skunk and Zorro the fox, Cooney was a great pet. If I held up a marshmallow out of reach, Cooney would climb up my leg, perch on my shoulder and eat out of my hand. He never bit me and I would play hard with him, wrestling and teasing him mercilessly to get him growling. We were like two litter mates grappling and mock fighting, but I like to think he knew we were just having fun. When fall came I was worried he'd get trapped or shot. Fur pelts were becoming valuable, with a good raccoon pelt in the 1960s bringing up to three dollars.

One afternoon Cooney came home with what looked like a bad dog bite on his back leg. I cleaned it up and hoped for the best. He recovered and then one day I realized it had been several days since I had seen him. I worried off and on all winter, hoping he was hibernating nearby, and then, the next spring I found him eating with the cats and we resumed our friendship. Sometime that summer he did move on for good, presumably to establish his own territory.

My memories of those creek roaming days are surrounded by a warm glow that is probably part nostalgia for the small farm lifestyle with its close ties to weather, seasons, and nature. Looking back at how much I learned about animal behavior from my wild pets, and how they cemented my love for wildlife, I feel a bit sad that young people today can't have a wild animal as a pet, though the animals I'm sure, don't share my sentiments.

Skunks on Horseback

My experience with skunks was not limited to Flower. One night when I was 13, my sister, brother, and some neighbor kids and I were on a horseback ride. It was a warm, moonlit night and we were cantering across a recently mowed hayfield. In the moonlight I saw the white back stripes of a female skunk with three kits silhouetted against the dark ground. I was immediately overcome with desire for one of those little kits, so I jumped off my horse, dropped the reins and sprinted up to them, throwing my leather jacket over the closest kit. To my utter disappointment it slipped out from under the jacket and into some tall grass and escaped. I

had my arms in the jacket sleeves before it sank in that all skunks are armed and dangerous after five to six weeks of age. I smelled so bad I had a hard time catching my horse and remounting. He didn't like my pungent smell and shied away several times before he let me mount.

My jacket was thick brown cowhide with a sheep's wool collar and resembled a World War II flight jacket. I had ordered it from the Montgomery Ward catalog with $20 from my saved-up lamb and wool money. A few years earlier, Grandpa Marshall had given each of us three kids a ewe sheep and twin lambs to start our own flock. As the flock grew larger each year we sold the castrated male lambs and sheared and sold the wool which became our lamb and wool fund. One of the wool buyers was Groenewold Wool and Fur Company. At nearly 70 years of age, I still sell some of my beaver, mink, coyote, and muskrat pelts to Groenewold.

That night when I made my odiferous way home I was feeling bad. Not only had I missed out on a skunk kit, but I feared I had ruined my jacket as well. When I got home, Mom said, "Take off your clothes, I'll get out the tomato juice." Tomato juice was the best remedy available back then to remove the scent of skunk from clothes, dogs, or anything else a skunk could spray. Today we use a mixture of baking soda, laundry detergent and hot water. Somehow the chemicals or acids in tomato juice usually did the job but my jacket was another story. The skunk left a stain on the inside lining, and it took many treatments and several months of airing out before I could wear it in public. To add insult to injury, by the time it was wearable again, I'd almost outgrown it.

Nature's Engineers, Friend or Foe?

"Choose a job you love and you will never have to work another day of your life."

—Confucius

My mother told this story about my high school graduation in 1969. After the ceremony, a group of friends and neighbors were invited to the farm for the coffee and cake, which was customary back then. I hung around the living room for a while and greeted everyone, but like most 18-year-olds I was aching to celebrate with my peers. So, after what I deemed a decent show of visiting, I slipped out, leaped in my car and was off to go to other parties and hang out with my friends. After I left, Mom said talk among the adults turned to the obvious subject, "What's Jeff going to do now?" Mom replied, "Well, we don't think he's interested in farming. Jeff is interested in a career in wildlife conservation."

Two of the men in the group—the sheriff, who was a neighbor, and the local veterinarian—looked at each other and said, "Yeah, we figured so."

"What do you mean," Mom said, "how did you know?"

Sheriff Grant spoke first. "Every night after school I see him tramping across the fields and sloughs with his shotgun hunting pheasants and rabbits. Then in the morning he's back out there heading for the creek with a pack basket of muskrat traps."

Doc spoke next. "I do vet work on over a hundred farms and nowhere other than here do I pull into the yard and see tame raccoons sitting on the porch or red fox running around the yard, let alone a kid walking a de-scented skunk on a leash in the fourth of July parade. Look around here, he's got cages full of rabbits, a chicken coop with over five species of pigeons, and then there's the turtles, snakes…you name it. Look at the cattle water tank—it's got channel catfish in it. Then in the fall the barn's full of muskrat, fox, and raccoon pelts. He spends all his time hunting, fishing, and trapping. If he goes to college what did you think he was going to major in, accounting?"

"Oh, I see your point," Mom said.

My First Dream Job

Well, Sherrif Grant was right. Fast forward six years to 1976. I was 25 years old, married to my high school sweetheart Sheilah, and baby Zach was on the way. I was just finishing college in northern Wisconsin and working a seasonal job at Brule River State Forest when I got a call about a permanent job with the Wisconsin Department of Natural Resources (DNR), Bureau of Wildlife Management. The job title was wildlife technician. It would be stationed at Park Falls, Wisconsin, a nearby paper mill town home to about 3,000. I thought about the position. While it was not as prestigious as a wildlife manager, I decided the technician position had a quality that was more important to me than money or status: adventure in the field! I applied, was interviewed, and to my delight, I was offered the job and I immediately accepted.

I started work that spring and learned that my work area covered Ashland, Iron, and Price counties, which contained substantial tracts of land including county forests, national forest, and a number of state wildlife management areas. The region's aspen forests were plentiful after the clear-cutting and forest fires of the early 1900s, and produced abundant populations of ruffed grouse and white-tailed deer as well as other game dependent on early successional forest types such as aspen.

Ducks, beaver, deer, and bear were central to the job and field projects often required a crew. That first spring when the deep snow pack and ice melted and creeks and rivers filled to capacity I learned how to adjust runoff levels on our flowages, reservoirs managed for waterfowl. It was an inspirational time to be out in the field, and I noted with pleasure migratory birds returning from the south, insects appearing and plants responding to the warming soils. Our crew loved hearing the honking of migrating V's of Canada geese and seeing the first mergansers land on an impoundment pond.

I soon learned that spring was also a time for upland gamebird work. Sharp-tailed grouse management gave me a first taste of prescribed burning. The spring, before "green up," was a perfect time to rid the sharp-tailed habitat of unwanted trees and shrubs but we had to plan carefully as it was also a season when wildfires could take off. I had to sandwich in gamebird breeding surveys whenever the weather allowed. The first of these was for woodcock, and it was a driving transect route with scheduled stops where I listened and recorded data. Woodcock surveys started at dusk, and entailed listening for the "peent" calls the males would give on their dancing grounds. If I was lucky and the birds were close to the road, I saw the mating flights and heard the twittering wing beats of these lusty little "timber doodles" who often peented under a full moon.

The other two species were grouse. Ruffed grouse required drumming surveys which I started at dawn, listening for males, who declared their love and lust by rhythmically beating their wings atop a fallen log. Sharp-tails required a previously constructed blind erected on a dancing grounds. I will never forget my first experience watching their breeding ritual. The assistant manager I worked for, Sam Moore, took me along to show me the ropes. Sam had been a professor at University of Wisconsin–Stevens Point before taking the job with DNR and I loved working with him. He had a wealth of knowledge and was one of my favorite mentors. I had woken at 3 a.m. and met Sam for the drive to the dancing ground in Price County. I was dressed heavily as Sam had warned me that it would be cold in the blind and we drank hot coffee as we drove. We were wide awake when we turned off the truck on a rural roadside a quarter mile from the dancing grounds. It was still dark and we quietly walked in and slipped into the blind. As dawn neared I heard the birds landing nearby with a soft whir of wings. At first light, the males began to strut and inflate their purple throat sacs. Things really got going when a female approached, and the males began cooing and quivering their wings. We sat recording data for an hour or so, until we thought we had a good count. About then, a red-tailed hawk flew over and flushed the birds, so we decided to slip out and head for the truck. By then I was shaking with cold and vowed to wear more clothes the next year. Little did I know that no matter how many layers of wool and wind block I added, I never would manage a sharp-tailed survey in comfort.

During that same time our crew conducted several weeks of deer population monitoring transects. These were quarter-mile walking surveys following a compass bearing with periodic stops to count the piles of deer "pellets" or droppings. The hundreds of data points we recorded were combined to estimate winter deer populations. At first, I couldn't figure out how the rest of the crew got done so much faster than I, until I discovered that they were giving me all the lowland swamp transects to struggle through while they had a "walk in the park" in nice upland woods.

That summer we battled biting flies and mosquitoes clearing one-acre grassy wildlife openings with bulldozers, maintained natural openings with herbicide, and seeded closed-out timber sale trails with clover. We responded to complaints with problem bear and beaver when necessary, and banded waterfowl. The waterfowl had to be captured at night using boats and spotlights—as close as we got to a chase scene. We operated bulldozers and tractors to maintain the dikes and trails on our management areas. My childhood hours running farm equipment came in handy.

Perhaps the biggest thrill for me that summer was learning and conducting the annual black bear survey. This involved packing 50 two-pound mesh bags of rancid meat and hanging them every half mile over a 25 mile transect. We suspended them in small trees or on overhanging limbs. The destructive claw marks were easily seen on the trees when the delicious morsels were devoured by a hungry bear. A week later we returned to the sites to check for activity. The crew explained that by comparing data over numerous years a population trend could be determined. The final check involved removing any bait bags bears had not hit, which by that time were odiferous and seething with maggots and carrion beetles.

If I had thought spring and summer were hectic, I soon learned that fall was the busiest, with hunting seasons and lots of time with the public. Harvesting black bear and white-tailed deer required registration, so we set up stations where we would collect harvest data such as age, sex and harvest location. I learned how we aged black bear by removing a small molar for cross-sectioning to determine age. I also learned deer aging, which required comparing a harvested deer's molar eruption and wear patterns to determine its age. I worked registration stations throughout the counties; these were often at popular bars, gas stations or sporting goods stores. I am outgoing and can tell a good yarn or listen to one, so that fall was exciting as I listened to hunting stories and used the harvest season as an opportunity to share our wildlife program goals, data and results with the public. After all, hunters who had just harvested an animal were in good moods and were willing listeners.

My worries that winter would mean too much time in the office were soon put to rest. During that season we conducted furbearer track surveys, monitored deer yard mortality, marked property lines on our state management areas, and, if time allowed, hand-sheared aspen to promote regeneration. I stayed in good shape as I logged miles on snowshoes.

Throughout these first months I was also learning another major part of my job: wildlife damage and nuisance control. Over the next few years I became skilled in abatement techniques, especially for beaver. One of my colleagues was Jack Koch, an expert on beaver control. Going out with him on a beaver complaint was one of my favorite tasks. In a typical scenario, a beaver colony would plug and block a road culvert, causing the water to flood the road. In the process, beaver would fell trees and the road base would become saturated, soften and wash out, creating a hazard to traffic. Even on rural roads the work of beaver can wreak havoc and we would get the call from the town or county clerk or road department supervisor to remove the beaver along with their inconvenient engineering. We had a farmer who couldn't get his cows to the milk barn because of a flooded creek and bridge, loggers who couldn't harvest timber due to wash outs, even major highways were blocked by water and trees. On occasion, beaver would plug stream outlets from lakes, which would back up water and flood cabin septic systems or basements. Beaver also loved to munch on ornamental trees in people's yards. Species like paper birch were a particular favorite. In perhaps the worst case I worked on, a beaver dam washed out a section of railroad track and when the train came through, eight gondola cars full of logs derailed and flipped over. Luckily no one was killed or injured.

Whatever the problem, our technique was to trap the beaver and sometimes remove their dams. Beaver fur was a valuable resource in the 1970s, with a large pelt selling for $30 at the fur auctions. For comparison, my hourly salary was $3.66, so a beaver pelt could be worth over a day's wage. The trapping season started in December and closed in March. Trappers had a catch limit of fifty beaver. Traps had to be set at least fifteen feet away from the lodge and dam, and harvested beaver had to be tagged and counted by a warden before the furs could be sold. Trapping was a viable livelihood for many people during the winter months, so rather than kill beavers in a complaint area, we did our best to live trap and relocate them to a remote area where they wouldn't cause further problems.

The trap we used was called the Bailey Beaver Trap and was made in Tomahawk, Wisconsin. The "Bailey" had two big clamshell-like jaws made of mesh wire, so that when a beaver swam over and bumped its T-bar-shaped trigger, the release of the trigger would cause the jaws to spring shut and the beaver would be caught inside unharmed. The trap was designed to be set in about six inches of water to conceal it from the beaver. It was big, clumsy, and awkward to carry, especially with a forty-pound beaver in it; but Jack was an expert with them and he taught me all the tricks. Jack even invented his own trigger modification that reduced misfires and improved trap efficiency. We did really well with the Baileys, but if we had a trap-shy beaver that had learned how to avoid the Bailey we would try foot-hold traps on a quick one-way drowning cable or the instant-kill body-grip Conibear style traps. A last resort, which I only used twice, was to sit out at night with a shotgun and flashlight guarding a hole in the dam. Beaver can't abide the sound of running water escaping their carefully constructed dam, so if we ripped a hole in the dam and allowed water to escape, it was likely that about dusk the resident beaver would arrive to repair the breach. That is when a carefully placed shot might be used to kill the animal, but it was an option we tried our best to avoid.

If we were trying to clear a plugged culvert, Jack had a "dynamite" trick. He'd cut about four inches off of a stick of sixty-percent nitro dynamite, and attach detonator cord, cap, and fuse, creating a super-sized firecracker. We'd then bury it in the mouth of a plugged culvert and when we set it off it would blow the culvert out like a charge through a shotgun barrel. The skill came in estimating the correct amount of dynamite. If we overestimated, it would blow the end off the culvert. This did happen a couple of times, but the town chairmen were surprisingly easy on us, perhaps glad to have their town roads open again even if it meant installing a new culvert. The worst culverts to deal with were constructed of concrete, which would crack in a blast so we couldn't use dynamite. I remember one concrete culvert in Price County that was about four feet tall and packed tight on the upstream side with beaver sticks and mud to the middle of the culvert. That day, Jack and I were working with Ralph Peterson, another wildlife tech stationed at Park Falls. Jack had me tackle the downstream side by entering the culvert to remove sticks by hand. Suddenly, the whole plug came loose and shot out of the culvert with a wall of water behind it. All I could do was ride the plug downstream with my clumsy waders tangled in the mess. When I reached the downstream edge of the culvert, Ralph was waiting for me, grabbed me by the shoulders and lifted me straight up as the mass of sticks and mud shot past. He'd timed it perfectly.

It was illegal and dangerous to blast alone, so being low in the pecking order I was usually along for the blasting. A typical large dam might take eight or more sticks of dynamite and sometimes there would be a series of active dams maintained by one colony of beaver. Blasting days often rivaled Fourth of July festivals in terms of noise, and to my mind were a lot more exciting. I apprenticed under Jack for a year, as required by law, and then took the test and got my own blasting license.

Leroy's Road

One time, Jack and I had a series of dams flooding a logging operation up on the Underwood Grade in Iron County. Roads were sometimes called a "grade" when they had been originally built as a narrow-gauge railroad grade during the logging era. When we arrived to investigate the site we suspected the beaver were gone, probably trapped out by fur trappers the winter before. The dams were holding water but there was no fresh mud on the lodge or dam and the dams had a few leaks. Before we went to the trouble of setting traps, we decided to throw a few fresh aspen branches into the pond and return to check a few days later when we had our dynamite equipment with us. Fresh green aspen is like candy to a beaver, especially after a long winter of gnawing on stale old sticks in their underwater food cache, so it was a good measure of whether beaver were present. When we came back, the aspen bait hadn't been touched, so we decided we could start blasting on the lower dams. The ponds behind these dams were small and each time we blasted we sent a supply of water downstream. We had worked our way up through at least three of these smaller dams when we finally reached the top dam. It was an impressive site with a big, deep pond behind a dam that was at least six feet tall, probably holding over three acres of water. In cases like this we liked to blast a small hole and lower the water slowly to prevent flooding problems downstream. In this situation, however, we weren't too concerned because the nearest road was over a half mile downstream and the water would have to pass through the series of coffer dams we had blasted earlier.

We punched a hole in the corner of the dam, placed three sticks of dynamite, and lit the fuse. The charge exploded with a boom and we were pleased—the blast had created a hole about three feet wide and the pond seemed to be draining slowly. Then, as we watched, a whole wing of the dam suddenly weakened and the remaining wall gave way. A tsunami of water gushed past us. Foolishly, we were not alarmed; the pond continued to drain and we thought everything looked fine.

When the pond was well-drained we packed up and headed out. It was a Friday night, we were already past quitting time, and we still had an hour's drive back to headquarters in Park Falls—

but just to be safe, we thought we should check the road. As we pulled up to the downstream culvert we couldn't believe what we saw—or couldn't see—the road was gone. There was a cut, six feet wide and four feet deep, in the middle of the town road and the culvert was washed several feet below the grade. I was scratching my head over what could have gone wrong, but Jack figured it out right away. The culvert was shiny and brand new, and had been replaced recently. The dirt and gravel around it hadn't been there long enough to pack in solid. When the wall of water hit, it easily washed out.

We worried that if a car came along it could fall into the hole, so we took some orange plastic flagging and tied it to tall sticks and jammed them into the hole to serve as a warning. Our other fear was that with ruffed grouse season opening the following morning, someone could be at one of the hunting camps up the dead-end grade and we might have blocked them in. We had no radio or cell phone in those days, so we drove off to find a phone. A few miles down the road we came to a lake resort and used their phone to call the Oma town chairman. I gladly let Jack handle explaining the predicament. The chairman said he'd call Leroy, the town road maintenance man, and get right on it—and we should go back to the breach and wait for him (our Friday night was shot for sure).

When we got to the washout, wouldn't you know it, there sat a guy in a jeep on the other side, and he was not happy. He was an owner of one of the hunting camps, but when we told him Leroy was on the way he turned around and left. About an hour later Leroy drove up with a dump truck full of gravel, pulling a trailer with a backhoe tractor on it. We apologized several times but Leroy didn't say much. We volunteered to help but Leroy told us to head on back to Park Falls, he didn't need our help, he could handle it. We were puzzled but more than happy to comply.

Thirty years later, when I was retired and living in Mercer, I heard someone at Tom's Cafe addressed as Leroy. His face looked familiar so I turned to him and asked, "Are you by any chance the same Leroy who worked on the road crew for the town of Oma back in the 70s?"

"Yes," he said, "That's me."

"Do you remember when two guys from the DNR washed out your Underwood Grade road on a Friday," I asked him, "the day before the opening of grouse season?"

"Oh yes," he chuckled. "I do remember that. I'd already put in my 40 hours so it was all overtime for me. No problem."

Unusual Passengers

One morning, Jack approached me and said something had come up. "I have to haul our bulldozer to Pershing Wildlife Area in Taylor County today. They need to do some emergency dike repair. Can you run the beaver trapline alone?"

"Sure," I said. "Let's see, we've got five Baileys set, and with that new complaint called in last night, I'll take two more empty traps and hit the road."

Jack said, "Great, I'll help you load up. Take my truck so you'll have four-wheel drive."

Now Jack's truck was a 1974 Chevy 4x4 short-box with a bumper winch mounted on the front. I loved it. It really looked cool, and Jack thought so too, as it was probably the only state truck ever to have gotten washed and waxed regularly. After we got the two Baileys loaded, there wasn't a lot of extra room for the cages we used to transport the live beaver to the relocation sites where they wouldn't cause trouble. We decided that with only five Baileys set, the odds were that I would catch three beaver at the most. We never had 100% success with the Baileys, and two Baileys and three cages perfectly filled up the truck box.

Standard protocol was if a beaver was caught in a Bailey, we would turn it on its side and squish the mesh down, compressing the beaver so one of its ears stuck out through the mesh. Then while kneeling on the mesh we could pull out the tag clippers and grab the beaver's ear and clip a numbered tag to it. This tagging was part of a study looking into whether translocated beaver stayed in the area where they were released, dispersed to new areas, or tried to return home. Eventually, we learned that of the dozens of beaver Jack and I translocated, all the tag returns from fur trappers were near the release site; no beaver had returned to the original locations. Of 272 beaver transplanted to lakes and potholes, their mean straight-line distance after being released was 2.0 miles.

After the beaver was ear-tagged we would transfer it to the cage and then transport the cage to the truck. Other than a lot of huffing and some teeth clicking, beaver usually settled down and were quite docile in a cage or trap. Traps are terrible for transporting beaver as the beaver can bite you through the large mesh. The bulky Baileys could also catch on the brush and sometimes fall over in the truck allowing the safety clips to come apart and the beaver to escape.

So off I went with Jack's truck, loaded with two traps and three cages. My first stop was at the new site, where I set the two traps. My next stop was at a damage site with two already-set traps, and much to my surprise we'd caught two beaver. I transferred them to two cages and headed down the road. My third stop, another damage site with two more traps, had two more beaver—so I filled my remaining cage but had an extra beaver. I looked around. The test holes we'd created in the dams the day before had been repaired so I needed to reset the Bailey traps, I was in a bind.

I sat down and thought for a bit. I couldn't transport a beaver loose in the truck bed because it could easily jump out. What if it rode in the cab with me? This might be the point where a person with a normal fear of wildlife would take a common sense approach, but caution, aka common sense, has never been my forte. I took the smallest beaver, which weighed about 30 pounds, picked it up by the tail, and tossed it into the cab and slammed the door. I quickly reset the Bailey trap and ran back to the truck. The beaver had jammed itself half under the seat but was too big to get further underneath. I slipped quickly into the driver's seat and started down the road.

The beaver did some scratching but I didn't think it could damage the vinyl seats or the floor of the truck. I had about an hour's drive to the next site and after about 10 minutes the beaver backed out from under the seat and hissed at me. I started talking to it in a soft voice, kind of like you would talk to a nervous horse. The beaver appeared to respond and started to relax and look around inquisitively. All this time I was clutching and shifting gears right next to the beaver, and trying to keep my eyes on the road and the beaver at the same time. To my horror, after a few miles, out of the corner of my eye I saw him start to nibble on the vinyl seat.

In a deep voice I shouted, "NO!" like I was scolding a dog. He stopped and obeyed. I started the quiet talking again and to my amazement, he pulled himself up on the seat and looked out the window. Now we were having a great time. There I was driving down the road with the beaver sitting next to me, staring out the window, moving his head from side to side enjoying the countryside. I pulled up to the last trap site, and you guessed it, there was another beaver—five out of five for a 100% trap success rate. We'd never seen that before and we never did again. This beaver was a little smaller, about 20 pounds, and after I tagged him, I just opened the truck door and tossed him in.

I had only driven a short distance when the fight began, with both beaver hissing and false charging each other and causing quite a ruckus. The first beaver stayed on the seat and the new beaver tried getting under the seat and even up under the dash. I started yelling again and they both froze. They must have figured I was the alpha beaver because they established a truce of sorts with one on the bench seat and the other on the floor. I only had about 20 miles to go before I got to the release site on the Flambeau River. When we got there, I pulled up close to the water, turned off the truck, and went around and opened the passenger door. Beaver number two jumped right out and waddled straight for the river. Then beaver number one, on the seat, carefully backed down onto the floorboard, peered out the door, slid to the ground, and followed his comrade to the water. The other three beaver in the back of the truck were easily released from their cages.

It was then, when I got back into the truck cab, that the smell hit me. The scientific name for beaver is *Castor canadensis*. The French name for beaver is castor with the accent on the last syllable. They are named for two special glands called the castors, which are located at the base of their tail. They also have two oil sacs located next to the castors. They deposit secretions from the castor glands and mark their home turf by constructing little mud mounds throughout their territory and squeezing some castor on each one. This warns other beaver to stay out, or be prepared for a fight. Beaver also secrete the oil from their oil sacs and comb it into their fur to repel water. I always saved the oil and castor from my winter trapline, to dry and grind to make a scent lure to use each season. The lure, along with some fresh aspen sticks placed in front of traps would ideally entice the beaver to swim into the trap. Castor scent from a foreign colony really gets beaver riled up and can be used to induce them to climb up on that mud mound and deposit some of their own castor. Beaver castor is quite valuable, even today. At one time in the 90s it sold for close to 100 dollars per pound—for trapping lure, but also as an agent in expensive fragrances. The perfume industry breaks down the chemicals in castor and extracts the ingredients that give perfume its staying power. The castor extracts are combined with flower oils and then sold in fancy glass bottles on the shelves of salons—in, for instance, Revlon's Chanel No. 9. Google beaver castor you'll find it used in dozens of products including cigarette tobacco and ice cream.

The castor chemicals were working well that day, because Jack's truck reeked of musky beaver. When I pulled into the Park Falls DNR shop, it was after working hours so everyone had gone home.

I went inside, got a five-gallon bucket and mixed up a cocktail of Pine-Sol, dish soap, toilet bowl cleaner and bleach. I grabbed a scrub brush and went to work. Thankfully trucks didn't have carpet in those days and the floor boards slanted towards the doors, so I put a spray nozzle on the water hose and went right inside the cab and gave it the works. After I was satisfied that most of the odor was gone, I found some rags and dried everything, then parked it with the windows down and went home exhausted. The next morning, Jack asked how everything had gone yesterday. I said, "Great, we caught five beaver." He never asked for details and I never provided them. And he never mentioned anything about the truck. It smelled great.

Loss of Affection

The call came one morning, shortly after I arrived at work, and John Olson picked up the phone. John was my boss at the tiny Ranger Station in Mercer where I had transferred in 1980. I was the wildlife technician, John was the wildlife manager, and our management area covered Ashland and Iron Counties. The call was from Jerry Best, the county DNR forester stationed 30 miles to the west in the town of Glidden. He asked, "Did Jeff have a beaver trap set on Willard Creek out on County Road N?"

"Yes," John said, "why?"

"Well," Jerry said, "he didn't catch a beaver, he caught a trout fisherman by the foot and the guy is really mad. Tell Jeff he'd better get over here fast. I'll let you talk to Marion, she's got the details, her father found the guy and let him out of the trap."

At this point, John gave me a nod so I picked up the phone extension and listened in.

Marion Pohl was the county forest department secretary. After a brief conversation, I got in my truck and headed for Glidden.

Now, I'm getting ahead of myself. This story really starts with a memo we'd gotten several months earlier from the district headquarters in Spooner stating that we (the DNR) were out of the beaver business. The higher-ups in Madison had decided that beaver control work was too costly and we were spending too much time handling beaver complaints. District staff felt time would be better spent working on bear control, habitat projects on the county forest, private land management, and wildlife surveys. They'd already put the hold on live-trapping and translocating beaver the year before, as live traps had to be checked daily and if no beaver were caught that meant a wasted trip, translating to loss of time and money. During my last few years of state beaver trapping we had quit using the live-catch Baileys and had gone to body-grip Conibear traps and foot traps on a quick drowning slide lock system. This method was more efficient as the beaver were killed quickly, humanely, were out of sight, and traps could be checked every three days, saving time and funding. Then the word came down, no trapping at all and we were just to assist by giving the public verbal advice. John and I fought it, arguing that many of these beaver complaints were serious. We'd even had trees felled over State Highway 51, one of the major north-south travel routes across northern Wisconsin. The DNR staff in charge didn't agree, and our beaver budget was removed to the dismay of highway departments, township road crews, and unhappy citizens who were left to deal with the fallout. The DNR had effectively transferred the work from one government agency to another, but the taxpayer would still pay. Costs were transferred to the counties and townships but the state was able to say it had reduced costs for itself and that made the bureaucrats and politicians look good.

John and I thought the change was illogical, but our hands were tied. In an effort to ease the transition, we put together a pamphlet on beaver control and abatement and offered some hands-on trapper education for town and county road crews. We got the okay from the district staff and sent letters to the highway departments, railroads, and townships inviting them to training sessions.

One of the first municipalities to respond was the Township of Morse. They had a complaint site on Willard Creek, east of Glidden, and their road crew said they would love to meet me there for a demonstration. I proudly handed them a copy of our new beaver control publication. I explained the equipment they would need: chest waders, a culvert cleaning hook, and traps. I told them about castor lure and gave them a bottle. I brought out a 330 Conibear (body-grip trap) and showed them how to set it safely. The site had a plugged culvert so I unplugged it, and as the water receded, I pointed out the spider web of channels the beaver had created leading into the tag alder shrubs along the bank. I picked a channel about a foot deep and a foot wide where the ten-by-ten-inch body grip trap fit perfectly. I placed fresh aspen branches (bait) behind the trap and a mud pie with castor lure next to the branch. I made a second similar set about 10 feet away on another channel. I told them to check the traps in a few days and let me know how they did. Then I got in my truck and went home. Apparently, someone else decided to go trout fishing later that day.

It takes about an hour to drive the winding roads between Mercer and Glidden. When I got there I went straight to the forestry office to talk to Marion, who filled me in on the details of the trapped fisherman. Her father, Ralph, who lived on County N, was driving to town, she said, when he saw a guy in chest waders sitting along the road frantically waving his hands in the air. Ralph pulled over and walked up to the excited fellow, only to see that he had a 330 Conibear body-grip trap clamped onto his foot. Ralph, being a beaver trapper himself, quickly compressed the springs, set the safety hooks and slipped the trap off. The fisherman, I'll call him Nils Johnson, jumped up and walked around a couple of minutes, shook his foot a couple of times and Ralph thought he seemed okay. His foot was a little swollen and had a red mark where the trap jaw had closed on it. Luckily, his chest wader boot was thick rubber and 330 traps have about a two-inch gap when closed. Ralph then read the trap identification tag, "Property of WI-DNR," looked at the pile of beaver stick debris cleaned out and lying next to the culvert and figured things out. Mr. Johnson then decided to resume his trout fishing and Ralph proceeded on to town.

When Marion had finished, I was flabbergasted. I believed her story, but I thought what a fluke—I'd never heard of this happening before. Yes, this trap has a ten-inch jaw spread that a boot would fit inside, but it is set vertically and chances that someone's foot would hit the trap at the correct angle to set it off and allow a good grip were astronomical.

Marion told me that Mr. Johnson was about 50 years old, and living at his mother's house because he was going through a divorce. I drove to the house, and with a bit of dread I went to the door and knocked. An elderly woman answered and I explained who I was and that I had come immediately after hearing her son had been caught in the trap. I expressed my dismay, asked if he was all right, and whether I could talk to him. She said I'd just missed him as he'd been out berry picking in the woods that morning, filled up all his containers and come home for more before heading back to the berry patch. It was a bumper year for raspberries, she said. She thought he would be home in a couple of hours, and suggested I stop back then. I retreated to the truck and headed back to the county forestry office where I took a lengthy lunch break waiting out my two hours.

When I returned, Mr. Johnson was home. He seemed like a pleasant fellow, happily chatting with me about the berry crop and trout fishing. He showed me his leg and when I saw the red mark, I told him I'd feel better if he had a doctor check it. He told me the soreness was gone and handed me back the trap. I headed out to Willard creek and pulled the other trap.

Weeks later the phone rang and when John Olson answered it, he glanced at me and then handed over the phone saying, "It's Madison, the state's legal department."

"Is this Jeff Wilson?" the attorney asked. I said it was. "Well I've got some bad news for you. You're being sued by a trout fisherman you caught in a beaver trap. He wants 50,000 dollars for pain and suffering the trap caused him."

"Oh no," I said, quite at a loss for further words.

"Are you sitting down, Jeff?" he said. When I replied in the affirmative, he told me, "His wife is suing you too."

"His wife, what's she got to do with it?" I asked, my voice rising a bit in astonishment.

"Well she's suing you for 10,000 dollars for 'loss of affection.'"

"Loss of affection, what's that?" I responded with some puzzlement.

Well, the attorney said, "It's kind of like this. When the wife says to her husband, 'How about we do a little romancing tonight?' and he says, 'Sorry, not tonight honey, I have a foot ache.'"

After I recovered my composure, I described the entire series of events in more detail and the DNR attorney told me to document everything, including time, dates, conversations and any witnesses. There would be a hearing before a judge in Ashland County Court and he would represent me and even though they were suing me personally, because I was acting as a state employee, if we lost or settled out of court the state would pay the money, not me.

Jerry Best called a few days later. "You'd better take a look at the Park Falls Herald's second page," he said. When I worriedly asked why, he read me the headline: "Trout Fisherman Caught in Trap." Three days later Jerry called again somewhat gleefully expanding on the theme. "Now, you're really getting famous, you've made the Chicago Tribune. The Associated Press picked up the story, so who knows how many papers you're in now." I could tell Jerry was having a really good time with this.

When I got copies of the papers it seemed that both had interviewed Mr. Johnson. The articles described how he had accidently stepped into the deadly beaver trap and agonized for hours, not only from the extreme pain in his foot, but also at the mercy of hordes of mosquitoes and black flies. The articles went on to describe how lucky he was to be alive today, rescued by a good Samaritan and freed from his pain, but that he might suffer forever from his permanent injuries.

Soon enough I got the full story from the local grapevine, this time from my friends at the county forestry office. Jerry and Marion told me that after I'd visited and talked with Mr. Johnson he had stopped in at the local pub in downtown Glidden and told his story. His buddies at the bar soon convinced him he had won the lottery; all he had to do was sue the state for pain and injury. They sent him to see Price County Attorney, Art Debardelaben, who was well known for his anti-DNR opinions. The guys at the bar thought Johnson could hire him on a percentage basis so the suit would be virtually cost free. As the story grew, Johnson's estranged wife got wind of it and saw an angle for herself. They weren't divorced yet and $10,000 would come in real handy for her. She decided to work with the same lawyer and added in the more personal elements of the suit, i.e. the sections relating to affection.

When word got around the DNR about my being sued by Johnson's estranged wife, someone said, "Gee, Jeff, if you're being sued for loss of affection maybe you should offer her some in replacement. It might save the state $10,000." I didn't think it was funny. I was worried!

The day of the hearing came and with it my apprehension grew. I met with the state's attorney in a room in the courthouse before the hearing. I told him how upset I was, and worried that this issue could cost the state $60,000, the equivalent of three years of my salary. He said, "I read your account and we've got a good defense. They have no medical records as Mr. Johnson didn't even go to the doctor until a week after the incident and at that time the doctor couldn't even see a mark. It's all hearsay based on Mr. Johnson's story so we're going to fight it and not settle out of court. Just go in there and tell your story."

I told the judge my story and submitted my account, including why I was training the township employees and how unhappy I was with DNR's policy changes concerning beaver control. He let me have my say and didn't interrupt me. At the end of the hearing, the judge looked at me and Mr. Johnson and dismissed both cases.

Thank You God

By the mid-1990s, the beaver population had exploded and harvest regulations had gone from a fifty-beaver limit to no limit at all during the trapping season. I was no longer assigned beaver control abatement unless there was a real emergency. The 15-foot trap placement setback that restricted trapping near and on beaver lodges was eliminated and the season was extended in both the fall and spring. Beaver pelt tagging by DNR wardens was no longer required. For a few years a beaver subsidy was paid to trappers in the spring to increase harvest. Early on it was 10 dollars each for beaver on trout streams, and then it dropped to seven dollars each for beaver anywhere in northern Wisconsin. But beaver still created "problems" in the minds of many people. Fish managers, especially those with a lot of marginal cold-water trout streams, disliked beaver and the impact they had on water temperatures and stream bank stability. Some fisheries staff thought the only good beaver on a trout stream was a dead beaver and groups like Trout Unlimited helped fund beaver removal.

The science shows that beaver dams on trout streams can be bad for trout. They warm the water by restricting the flow and spreading the waters out over broad shallow areas, sometimes so much that trout can't survive. Beaver activity reduces shade trees over streams, causing further warming that can cause a decline in dissolved oxygen. The spread and deepening of water in the pond can also saturate the overhanging original stream banks, causing them to cave in, and silt up the trout's gravel spawning areas. Fish managers complained bitterly that after the beaver left and the dam washed out the remaining streambanks were straight and left little cover for trout.

We in wildlife management had our side of the issue too. Beaver have coexisted with brook trout for thousands of years. Beaver ponds create valuable habitat for cool and warm water fish, waterfowl, osprey and mammals such as muskrat, mink, raccoon, and otter. Abandoned (drained) beaver ponds turn into sedge meadows with plants favored by deer and bear, and nesting habitat for songbirds. Frogs, turtles, and other amphibians and reptiles thrive in beaver ponds. I didn't think the wildlife bureau fought hard enough with our case for the beaver. The fish managers prevailed, and elimination of beaver on trout streams was still a priority.

One day the Mercer fish crew asked me if I could help them clear the beaver dams out of Alder Creek in northern Iron County. Both Jim Cox, the fish management technician, and I had blasting licenses. The beaver were about trapped out, but several dams still held water. The Mercer crew felt removing those dams might improve the stream for brook trout. We had about three miles of stream to clear and started early in the morning at the furthest downstream dam. There were four of us—myself, Jim Cox, Dennis Scholl, and Tom Cawst. It was a hot July day, and getting hotter by the hour, and we were wearing chest waders. We had a lot of tag alder, swamp, and mucky streambed to wallow through. After we blasted a hole in a dam, we used potato digging hooks to pull the dams open wider. It was hard work and we soon ran out of drinking water. By noon it was clear we had misjudged how long this project would take and how many dams were on the stream. It was late afternoon by the time we had made it to the top and blasted the last dam, so we sat down for a brief rest before tackling the half-mile walk out to the truck.

Someone said, "Man, am I thirsty." Another said, "I sure could use a cold beer right now." Then Jim said, "Look!" and right before our eyes four cold beers stuck in a plastic six-pack ring appeared, drifting and rolling in the current toward the hole in the dam. Someone scrambled to his feet and reached the beers, lifting them out of the water. I don't remember what brand they were, but they were cold and dripping. How could this be possible? Jim knew an old-timer named Grassley who fished this stream a lot. He suggested that perhaps Grassley had stashed them somewhere above the dam for his own use. We stared at them for a few moments. Four cans of cold beer. Four of us. But state employees are not allowed to consume alcoholic beverages while on the clock. We made a unanimous decision. We were out of water, so it was most certainly an emergency. We cracked them open and beer never tasted better. When we finished the beers, we looked up at the sky and said, "Thank you, God."

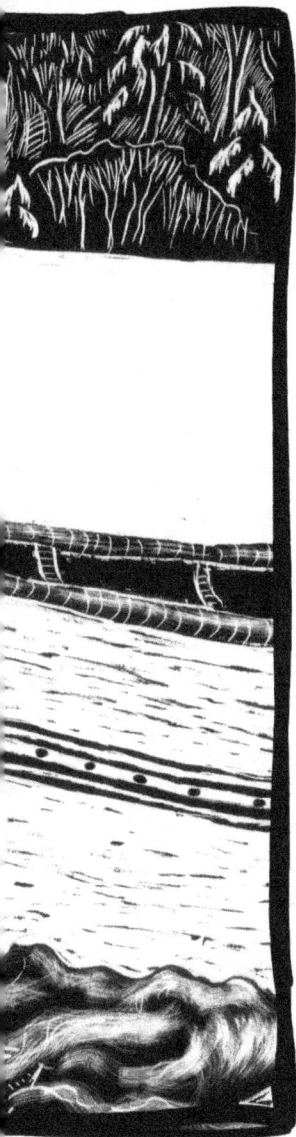

Darwin Awards

"There are three kinds of people in the world. Those who can count and those who can't."

—Author Unknown

As is probably already apparent, beaver were not the only animals imperiled by my adventures trapping. In a few instances, it was all I could do to save my own skin and I will relate some of these anecdotes here as evidence that beaver in Wisconsin are worthy adversaries. In fact, my family likens some of my adventures to the notorious Darwin Awards. For those of you who aren't familiar with this particular honor, a Darwin Award is given to people who have eliminated themselves from the gene pool by doing stupid things. I have not yet received a Darwin because, as I write this, I still have a pulse and to compete for the real award you must be deceased. You will likely recall from high school biology class that Charles Darwin's theory of evolution is rooted in the concept of survival of the fittest. The individual with the best skills for species survival manages to pass on their genetic material to offspring. The Darwin Award is a somewhat tongue-in-cheek spoof on those who should, perhaps, not pass on their genes.

Each year, I hear the Darwin Awards reported on National Public Radio and some stand out in my mind and may help you envision the criteria. One particularly dramatic story was about some Russian fellows who drank quite a bit of vodka. One inebriated fellow started a manly competition by picking up his chainsaw and saying, "Watch this!" He started the saw, ran it at full speed, swung it around and stopped it a fraction of an inch from his neck. One of his drinking buddies said "That's nothing, now watch me." The buddy then grabbed the saw, started it up again and swung it towards his own neck but misjudged and decapitated himself. He was the Darwin winner that year. Another classic winner was a man who climbed an electric pole, sat down on top of it, and drank a six-pack of beer. After the last beer had been imbibed he had a great need to empty his bladder so he decided to relieve himself. This was all fine and good, until his liquid stream hit a live wire. The electric current moved upstream, so to speak and bingo, another Darwin Award winner.

As I mentioned, I have not yet qualified, but my friends and family often remind me that I have come close to a nomination several times, so I am going to tell you the stories of my three attempts to date.

The first try at a Darwin occurred back in the late 1980s, when I was running a boat trapline on the Turtle Flambeau Flowage (TFF) near my home in Mercer. It was late fall and beaver prices were pretty good. I had scouted 13 colonies of beaver that fall, and most could be easily reached by boat. I would motor up to each colony and set Conibears on the lodge channels and foot traps along the shore. Once I had caught the larger adults, I'd pull my traps and move on, leaving the smaller beaver to maintain the colony. I also set raccoon and mink traps near the colonies as the shoreline ecotone provided by beaver meadows are favorite habitats for them.

I was just finishing up this trapline and had about 30 beaver pelts on stretcher boards when I heard a cold front was coming. I decided to play it smart and pull all my traps before the freeze-up. After a few hours on the water, I had the last of my equipment and catch stowed in the boat and headed for the landing. I was about two miles away when I hit something hard with my propeller, the outboard motor flew up out of the water and the motor killed. I was in a channel between two islands, so I rowed to shore to check if my propeller was bent and needed to be replaced.

Back in the 1980s, there were no accurate contour maps for the flowage and boats rarely had depth finders to warn you of shallow water, at least mine didn't. Because the flowage was created from 16 lakes flooded into one in 1926, unless you are over an old creek, river, or lake bed you can hit rock bars, stump fields from the old flooded forests, and boulders the size of Volkswagen Beetles barely under the surface. Also, in the fall, just before ice up, the flowage is at "full pool," which causes logs along shore to float around, sometimes just under the surface. These logs are called deadheads.

My outboard motor was a 25-horsepower Evinrude with tiller steering and I had hit plenty of objects with it in the past. When I got to shore, I tilted the motor and looked over the propeller. It was then that I saw the gear oil leaking out of a crack in the gear housing above the prop. I thought I might have hit a deadhead and it looked like my luck had run out. This motor had hit one too many objects in its day. If I started the motor and put it in gear the pressure would pump out all the oil, which would damage the gears, causing a very expensive repair bill.

The boat was 16-foot Grumman, with a deep V-keel and steep sides. I had oars, but that boat was not a good rower and I was two miles from the landing. On top of that, the boat was full, and I would have to row into a headwind. There were no cell phones in those days to call someone for help.

As I was scratching my head about what to do, I heard a boat coming just beyond the next turn along the shore. A moment later, I could see two young duck hunters in a canoe with a two-horse-power motor on the back. I waved them down and explained my predicament. They said they were going to the same landing where I was parked and offered to give me a ride. They were pretty loaded down, so I didn't even consider asking for a tow considering my heavy boat and the distance and wind. Boy, I thought, what luck just to get a ride out of here.

The next day I headed back to retrieve my disabled boat. I took my canoe with a side-mounted four-horsepower motor. The motor was also an Evinrude, the same brand as my disabled motor, so my plan was to mount the smaller motor alongside the 25, hook up the same gas line and then motor back to the landing, towing the canoe.

Everything went well until I got to the bay where I had left my beached boat. The cold front had come through and the bay was frozen over with about an inch of ice. It was about 100 yards to open water from shore, so I landed the canoe up the beach where the ice was thin and carried the four-horse on my shoulder to the boat. I grabbed my small Hudson Bay ax, chopping the boat free from the ice; finally opening up a big enough space to turn it around and point toward open water, while staying warm and dry in my chest waders. I attached the little four-horse alongside the damaged 25hp, hooked up the gas line, and fired it up. The ice was thinner near the warmer land mass so for the first 20 yards or so the motor easily pushed the boat, the bow breaking the thin ice. Every now and then, however, those churned up chunks of ice would work their way back and hit the propeller, killing the motor. Then I would have to clear away the ice, restart the motor, and proceed on through the ice. As I got farther from shore the ice started to thicken and we slowed down until it was so thick that the motor could no longer move the boat.

I was stuck. I turned the throttle on to full speed, tightened the steering in place and went to the bow of the boat, jumping up and down, making the bow create waves to crack the ice. This worked for a while and I made it another 20 yards or so before I was again stopped by the thickening ice. At this point I only had about 50 more yards to go before I would reach open water and I was growing impatient. My next move was to lean over the sides of the boat and chop at the ice with my Hudson Bay ax. This worked and I made it another 10 yards, but the boat had such tall sides that the little ax barely reached the ice. Well, I thought, why not see if I could break the ice with my foot? So, I straddled the gunwales and used one leg to stomp on the ice. It was faster than the ax method and my chest waders kept my foot dry. I started alternating, stomping first off the starboard and then off the port side of the boat. When a space was opened, the boat would lunge ahead a few feet and I would repeat the process. Once in a while, broken chunks of ice would work their way back, and hit the propeller, killing the engine. I would then have to clear the ice with an oar and restart the motor. Finally, I only had about 20 yards to go to reach open water. I was pretty hopeful until I hit a thicker stretch of ice, perhaps stronger due to lack of current. Stomping with one leg wasn't working, so I slipped both legs over the gunwale onto the ice, using my shoulder and elbow to hold my torso alongside the boat and stomped with both legs.

In a split second I was through the ice and barely grabbed the bow with both hands in time to keep from going under. As my body broke the ice, the boat took advantage of the opening and lunged forward, running over me. In a moment I was swung around, under the bow keel and pinned with the boat on top of me. The motor was running full throttle with the boat bow against my chest and my back pinned against the ice. I frantically tried to pull myself up by the gunwale, but each time I pulled and fell back, I broke ice with my back and the boat lunged forward pinning me tighter against the ice. During my lunges to get free, my chest waders had gone below the water line and filled up. I felt like I was inside a water balloon filled with ice water.

In that second, I realized I was in a dire situation, pinned against the ice and under the bow of the boat with ice on each side of me. I couldn't climb out, I couldn't move the boat. The motor was running full throttle pressing me against the ice!

I was just about at the point of believing that I'd bought the farm, as they say in Iowa, when a strange thing happened. My struggling must have broken off some chunks of ice and one or two had worked back and hit the propeller. The motor killed!

I knew this was my only chance, so I tried to pull myself into the boat chin-up style, but I was too heavy with my waders full of water. Instead, I pulled the bow down and the boat backed up. A couple of chin-up attempts later, and I had enough room to work myself hand-over-hand to the back of the boat where the sides were less steep, and with one great heave I flipped into the boat like a beaching walrus. I then pulled off my waders, dumped the water out of them, wrung out my socks and other clothing and put them back on again. I started up the motor, put it into reverse, and returned to shore. There I threw out the anchor, detached the four-horse, and headed back to the canoe. I was so mad at myself for getting into this predicament that I never felt cold, just angry. When I reached the truck, I put the heater on full blast and by the time I hit the paved road I was fine.

Two days later, I returned with a large boat borrowed from a friend, only to find the ice gone and my boat peacefully moored on the shore. The cold front was over, things had warmed up, and wind had blown the ice out.

After the shock of what had almost happened hit me, I sat down and had a talk with my ten-year-old son, Zach. I told him in the future, I would leave a note indicating where I was going and when I would return, with a list of names and phone numbers he should call if I didn't return on time.

The following week, I was returning from an outing and was about 20 minutes behind schedule. As I turned into the driveway, I saw a car with Joyce Olson, my boss's wife, at the wheel. She had just picked up Zach and was taking him to their house. She rolled down the window in her car and said, "Zach called and said you were late and probably drowned out on the flowage. I told him we were just leaving for Saturday night mass and we weren't going to look for you yet. I told him to put on some good clothes and I would take him to church with us. I told him he could pray for his numbskull father." She gave me a sly grin, clearly having heard the story of my dunk in the ice.

My next misadventure in survival of the unfit involved a winter trapping technique invented by a famous trapper from Hurst, Ontario named Paul Millette. The technique starts with clothing: you wear a pair of neoprene chest waders over your normal warm winter clothing so you can get in the water under the ice. The basic method is to approach a beaver lodge by snowmobile or on snowshoes, look it over and determine where the underwater entrance runs are located. You then cut a two-foot square hole over the runs with an ice chisel—or, if the ice is thick, a chainsaw—and then clear the ice away. Next you jump in the hole and hold yourself up with your elbows on the solid ice around the hole. While dangling there, you feel around with your feet until you've located the runs into the lodge. Sometimes it's shallow enough to stand on the bottom. But often it's deeper than the waders allow, so you use your toes to feel your way along the side of the lodge. If necessary, you cut additional holes until the runs are located. With practice you can get pretty good at this. After the initial foot survey, you heave yourself out and mark the exact center of the entrance in the snow next to the lodge. You then set a Conibear body-grip-style trap, attach wires to the side springs and lower it down under the ice directly in the center of the run. You can even set several, one in front of another, if it's a long run. You then cover the hole with snow, mark its location on the ice, and return in about a week to check the traps. This technique works well and is efficient and humane.

On the day in question, I pulled up to a large beaver lodge on the Turtle Flambeau Flowage. I got out my ice chisel and started chopping around the lodge, looking for a run. When I got to the southeast side of the lodge, my chisel went right through. Boy, I thought, that was easy.

I didn't stop to analyze things, but looking back, three things were working together to make the ice over this run so thin and easy to chop. First, it was March and the TFF is usually lowered a foot or two in March to make room for the anticipated run-off from snowmelt in April. Second, there were a lot of beaver in this lodge creating a lot of activity and heat under the ice where the run opened up. Third, the prevailing winds in winter are out of the northwest, and because this lodge was out in the open it had blocked the wind, creating a two-foot drift on its southeast side, effectively insulating the ice. These three factors created a unique situation on the southeast run—rather than firm ice over water, there was a space under the ice between the lower water level and the ice, and layers of crusted snow drift interspersed with very thin ice above the run.

I chiseled a two-foot hole, felt below with my chisel handle, and jumped in to do my foot survey, expecting to be held up by solid ice under my elbows and shoulders. But as soon as I had lowered myself three-quarters of the way with my elbows, the ice gave out on one side and down I crashed. I landed with my feet resting on a ten-inch thick underwater log sticking out of the lodge.

The top of my chest waders was about eight inches above water and it was at least 12 inches from my head to the top of the ice. I was completely hidden down in the hole. I was still holding my chisel, so I used its five-foot length to feel around below me and I soon realized there was no bottom to be found. This lodge was on the side of a bog and there was no way to know how deep the water was. I had to get out of there. My first thought was to use the chisel bar, shoving it into both sides of the hole above me and trying to do a chin up and pull myself up and out. The bar went in just fine, but when I pulled on it, the ice broke and the chisel came right back into my lap. It was at that moment that I realized it wasn't ice above me, just layers of crusted snow drift. It had somehow held me when I stood on it from above, but when I pulled on the bar from below it easily broke from my weight. I tried a few more times, hoping for a sturdier place for the bar, but every time I tried to pull myself up, the hole got bigger—until it was a larger diameter than the chisel bar. Also, I had only one perch and couldn't move from my spot, balanced as I was on the ten-inch log sticking out of the lodge. I tried to climb the side of the lodge and break myself out. That failed because the sides were vertical and the sticks were slimy and brittle, so I just stood there balancing on that log below the ice pondering on my dilemma.

I was warm and dry, and comfortable in a sense, but I just had no way to get out. My snowmobile was parked above me in plain sight. It was canary yellow and it stood out well. The problem was that this was the 1990s, so no cell phones, and Zach was away at college, and there was no Joyce Olson to be notified to send out the troops. To make matters worse, it was March: walleye fishing season had closed and the recreational snowmobiling season was about over, so the odds were poor that anyone would notice me and come over for a look. I can't remember if it was a Saturday or Sunday, but on Monday if I didn't show up for work perhaps someone would come looking for me. However, I was not sure I could last down there until Monday. I had to get out on my own.

As I worried over the problem, it occurred to me that the beaver had built this lodge with hundreds of sticks and poles. I just needed one strong enough to hold my weight and long enough to span the hole to good ice. I looked at my chisel. Where the chisel blade was welded to the handle bar was a little notch because the blade was wider than the bar. I started feeling and poking around in the water beneath me with that chisel. When I felt the chisel hit a pole, I tried to hook it with the notch and pull it to the surface. After about 20 minutes, I had pulled up several poles and sticks but they were either too short or too thin. I was somewhat limited in my abilities due to the balancing act I had to perform while standing on the log. After another few minutes of searching, I found a pole about three inches thick and about 10 feet long. It had no bark on it and looked strong enough to hold me. I could pull one end of it to the surface by hooking it on the chisel notch, but the other end led into the lodge about two feet below me. Try as I might, every time I got that one end to the surface, it would slide off the notch before I could grab it. I finally decided to try and cut it loose from the lodge with my chisel. I bent over, peering through the dark water and estimated about where it was attached to the lodge. I aimed my chisel and started striking. I'd hit it about one out of every four tries, but eventually I saw wood chips starting to float to the surface. After several minutes of chipping away, I hooked it again with my chisel notch, and this time when it came near the surface it broke and floated to me. I had my pole. It didn't take strength to chin myself up and out of that hole; I did it on pure adrenaline.

The last incident I'll describe was not so much life-threatening as humbling and painfully so. My first wife, Sheilah, used to say to me, "You men really don't know pain because you haven't experienced childbirth," and, naturally, I didn't argue the point…at the time, anyway.

Right below my cabin is a natural shallow pocket along the bank that every passing mink and raccoon investigates. Each trapping season, I dig a small hole in the back of it and bait it with some fish cleaning remains. I then set a 160 Conibear body-grip trap in front of the hole. A 160 is six inches square and makes a quick humane kill similar to how a common snap mousetrap works, except it has two jaws instead of one, and catches the targeted animal by the body or neck. I'd usually catch a mink or two and always a few raccoons. I'm not fond of raccoons, even though I had them as pets when I was a boy. Now I have a bit of a bias against them because they are the number one predator of common loon nests. A pair of loons usually nests in the bay in front of my cabin and I'm lucky if they bring off a hatch one out of every three years between the mink, raccoon, and eagles competing for their eggs.

Anyway, one day I pulled up to my cabin shore during trapping season and I noticed something had been messing with the trap set. I had baited it with the remains of a northern pike and some animal—a weasel, I presumed—had gone through the trap without setting it off, dragged out the bait, and dropped it right on the bottom jaws. Now I had made this set perfectly, hidden the jaws and springs and blended everything to look natural. I didn't want to go through the time-consuming job of tearing everything apart and remaking the set. I just wanted to put those slimy fish remains back in the rear of the hole. I picked up a small stick and tried poking the pike parts back in. This sometimes worked, but the pike was very slimy and kept slipping off the stick. So, I picked up another stick, and using the sticks sort of like two-handed chopsticks, got the last chunk moved back where it belonged. Unfortunately, when I was on my knees pulling out the sticks, I somehow bumped the trigger and whack: I was caught with parts of both hands, including my thumbs, inside the trap.

I've been accidently caught in plenty of traps—almost every trapper has. Usually, I could simply step on the springs and release myself if it was a foothold trap, or, if it was a Conibear like this one, I could release myself by compressing the springs and flipping on the safety catches with my other hand. It was the two "caught" thumbs that were the problem. There's a very good reason why we have thumbs and this was sure proof that humans would not have evolved very far without them.

Somehow, using my pinky finger and my teeth, I got the trap unwired from the tree root it was attached to. In my pain, my first thoughts were a bit muddled. I thought about the tong tool I owned, which is designed for compressing the springs of Conibear traps. I thought maybe I could work that with my knees. Then I thought, no luck, as the trap had two springs and after the first spring was compressed I'd need to flip over the safety hook to secure it before I could compress the second spring, and I couldn't figure out how I'd do that without thumbs. This line of thinking was not particularly helpful anyway, because I soon realized I didn't have my tongs with me.

By now, I'd been in the trap a couple of minutes and was starting to feel some real pain, so I struggled up the hill to my cabin. I opened the tool chest with my pinky finger and my shoulder and stared at the tools. I saw hammers, pry bars, vise-grips, and a hack saw and I tried to think what would get me out of this trap. None of these tools would work without thumbs. My gaze lit on the crowbar and I lifted it out with my pinky finger, then sat down and using my knees I tried jamming it between the jaws. I then applied leverage with my chin. I immediately realized that as I pried open one side of the jaw, the other jaw would squeeze tighter on my other hand. It was like putting my hand in a bench vise and cranking it down until I couldn't stand the pain anymore.

By now several minutes had passed and I was feeling much more pain and numbness. I was contemplating whether I should try to start the boat and find a fisherman to get me out of this trap. I wondered if I could even turn the ignition key or steer the tiller and whether I could twist the throttle to apply gas. I had a cell phone, but who would I call and how long would it take them to get here? It was three miles to the closest landing.

I finally went back to the toolbox and found a screwdriver and picked it up with my teeth. Using my wrists, knees, teeth, and elbows I managed to jam the handle between the spaced floorboards of the porch and push it between the trap jaws and just bend it against the jaws, literally pushing through the pain. Out came one finger, then another, then another, and then the thumb. With this free hand I quickly compressed both springs—and both hands were completely free of the trap. I squeezed and shook my hands as I walked in circles to regain circulation. Finally, I went inside the cabin and opened the medicine cabinet, grabbed the Tylenol and downed a double dose with a glass of water.

I returned to the porch, sat down in a rocking chair, nursing my hands, looked out over the bay and asked myself, really…just how painful is childbirth anyway?

The Scent of Skunks

"A boy who picks up a cat by the tail learns a lesson in life that can be taught in no other way."

—Mark Twain

Dancing in the Moonlight

Like so many crazy time-consuming things that happen, the call came at the end of the day on a Friday. When I picked up the receiver I heard a woman's voice: "My husband told me to call you, and we have a skunk in the basement."

"In the basement, how did it get there? If it's a walkout basement just open the door and it will eventually find its way out."

"It's not a walkout basement," she said tersely, "and as to how it got there, you have to talk to my husband about that."

As I got my gear together, I pondered the call. "Why is it always the wife who calls with the bad news?" I wondered. In my disgruntled state of mind, I speculated that it might be because men are usually too embarrassed to ask for help, and especially if the help has to be sought from the government, let alone the DNR. Coming from Iowa, where the public loved our natural resources agency, I'd been shocked by the attitude I'd encountered in Wisconsin, succinctly summed up in bumper stickers: "DNR— Do Nothing Right"…"DNR—Damned Near Russia" and "Support the DNR—with a Noose." The Wisconsin DNR is too big, or so the complaints went; they regulate too harshly, they won't let me do what I want with my land or my lakeshore and I pay high taxes. I should be able to do what I want with my land, after all, I own it. We don't need so much government. And yet, who do they call when they need help?

My truck was parked outside loaded with a tent, sleeping bags, and a cooler full of ice and frosty beers. This was not just any Friday. Tonight was the opening of the fourth annual Great Northern Bluegrass Festival in Mole Lake, Wisconsin, where I was planning to spend the weekend. Headlining were names like by Bill Monroe, the father of bluegrass music, and also appearing were Earl Scruggs (who wrote the theme for the Beverly Hillbillies TV show), and other big names like John Hartford and Benny Martin. In an hour I was supposed to meet a very nice, dark-haired, blue-eyed gal named Anna who claimed she could dance me into the ground. I had to work fast if I was going to solve the skunk problem and make my date. I got some bait out of the freezer, grabbed a cage trap, flashlight, and a gunnysack, then loaded the truck and drove to the complaint site.

"How did you get a skunk in your basement?" I asked the husband when he met me at the side of the driveway.

He said, "Come over here and I'll show you." We walked to the back of the house and there in the corner of the basement window was a softball-sized hole. "The kids hit a baseball last week and it broke the window. I haven't had time to fix it yet." It appeared that the skunk had somehow stuck its head in, sniffed around and fallen in, dropping eight feet to the basement floor.

"How do you know he's in there?" I asked.

"We got a whiff of his smell from upstairs. I turned a light on and saw him scamper under some shelves. He's down there all right."

"Is there any way out?" I asked.

"Just up the steps through the living room, and out the kitchen door. What are we going to do?" A hint of anxiety crept into his voice. "If he sprays down there, we'll have to move out. I've got a wife and three kids. We have no place to go."

"Well," I said, "here's what we're going to do. I have this cage trap. Skunks are easy to catch in these traps, they're not too smart. I'm going to bait this trap with some tasty morsels, then I'm going to wrap the trap in this gunnysack so only the door is exposed and carefully, quietly walk down those basement steps with the trap and place it on the floor in full view of the top of the steps so you can check it from upstairs. When the skunk gets caught, you'll see that the trap door is down. I'm going out of town for the weekend so when the skunk is caught just quietly walk down the steps, slowly pick up the trap and carefully walk up the stairs, through the house and out the door. Just so you know, it is illegal to release a live skunk in Wisconsin so once you catch the skunk you can dunk the skunk and trap in the lake or gas the animal under your car exhaust pipe. Oh, and one more thing, see this handle, it's in the middle of the top of the trap. You'll need to feel through the burlap sack and make sure you have a good grip on the handle. Don't drop it. If you do the skunk could spray.

"Oh, and one more thing, sometimes they chew up the gunnysack from the inside of the trap, scratch it up, and chew holes in it. You don't want the skunk to see you so get an old rug or another sack and if you think it's too chewed up take a broom handle, lay the rug over the handle and reach out with the handle and place more covering over the trap before you pick it up. Okay, you got it?"

He looked at me, his wife standing behind him listening, three kids huddled around her legs. "No way," he said. "That's crazy. I'm not going down there. If that skunk sprays we're finished. This is your job; you get paid to do it. Skunks can carry rabies...I'm not going near it."

My heart sank...but I thought a minute. My boss John Olson was out of town for the weekend and our student intern was gone as well. I considered asking the fire control crew to help out but discarded that idea; they would not appreciate the request. Then I thought of the game warden, weighing up the number of times I had assisted him and whether I could push for a favor. No, I thought, not for a skunk. In the ranger station work hierarchy asking for help on skunk control was really pushing the limit.

I looked at my watch. Anna had left Hurley by now and would be in Mercer in a half hour. What would I tell her? "I'm sorry I can't go this year. I have to catch a skunk and can't leave it in a trap in a basement all weekend." I started down the basement steps very slowly, trap in one hand and flashlight in the other.

When I reached the bottom, I carefully placed the trap in the middle of the room, adjusting the burlap and checking to make sure nothing was interfering with the trap's door mechanism. I was ready to head up the stairs when I heard a sound, a slight scuffling of feet near some upright garbage cans. I straightened up slowly and sized up the three long steps to the cans. I thought why not take a look, and quietly eased towards the cans. When I was close enough to see the dark void between the cans I flipped on my flashlight—and there he was, right below me between the cans with his head peering forward. I switched off the light. He didn't appear to know I was there.

I recalled my neighbor back in Iowa, Don Yanda, the expert skunk de-scenter who gave me my pet skunk Flower. At one point he'd told me that if you grab a skunk by the tail and lift it up off the ground, it can't spray you because a skunk has to have its back legs firmly on the ground to squeeze the sphincter muscle that shoots out the spray. "Could I just make a grab and lift this skunk out from between those cans?" I wondered, "What if I missed or dropped it and it sprayed the basement?" Anna, meanwhile, was probably just now pulling into Mercer. She was not going to be happy if I had to cancel.

I gritted my teeth, envisioned the motion, then reached down, made the swipe and grabbed that skunk by the tail, yanking it up and out from between those cans. I swung it around, took three steps and headed up that stairway, yelling, "Open the door, and clear the way." The basement door flew open, kids scattered, and I walked through the living room, hand outstretched. I continued through the kitchen and out the door with that skunk bobbing from side to side and his little feet spinning and trying to catch something he could push against. The family followed at a safe distance, and as soon as I got far enough away from the house, I gave that skunk an underhand softball-type pitch across the yard. He bounced twice then scuttled into the woods without even stopping to raise his tail.

The family stood there staring at me with their eyes wide. No one said a word. I said, "I'll be back for the trap Monday. Oh, and

I'd fix that window real soon if I were you." I jumped in my truck and drove away with a grin and high hopes for the weekend.

Sunday night when we got back to Mercer, Anna was curled up asleep against the passenger door. She'd been asleep for over an hour and while I was tired too, she had not danced me into the ground. With a full moon overhead I had danced her into the ground. On Monday morning I drove back to retrieve my trap from the homeowners. The cage was sitting on the sidewalk by the front door with the burlap bag neatly folded beside it. I looked over at the basement window and could see the glint of shiny new glass.

Skunk Soup

Once again, I was alone at the Mercer Ranger Station office when the call came in. When I picked up, a woman told me that there was a skunk in her yard with a soup can stuck on its head. She gave me the address, I jumped into a DNR truck, and I was there in less than five minutes.

She was right. In the middle of her neatly-mowed yard was a skunk walking in circles, its white stripe unmistakable. I could see the can label too, the colorful red and white lettering advertising Campbell's chicken noodle soup. It looked like someone had opened the can halfway with the can opener then pushed the cut edge inside the can. The skunk must have dug into a garbage sack, stuck his head inside the can to lick it clean, and got the cut can lid caught behind its ears. It was a younger skunk but full-size. I knew what to do: repeat the lesson learned in that Mercer basement. If the skunk couldn't see me with the can on its head, I could

pick it up by its tail, avoiding the risk you would normally have within 20 feet of a fully functioning skunk. I knew what the radius of their powerful spray was; I'd learned the hard way. So, I quietly snuck up behind the skunk, grabbed the tail and lifted it off the ground—then gently worked the can off its head, with no injury to the animal.

Looking back, I am not particularly proud of my next decision, but times were different then and standards of care for animals have come a long way. At the time I was pleased with myself for extracting the skunk safely and my devilish side got the better of me. I thought, "I have a captive skunk in hand, I've got to get some fun out of this." The truck I had was a Dodge with an automatic transmission. "Perfect," I thought, as I opened the driver side door and rolled down the window. I then got in the truck with my left hand sticking out of the window holding the skunk.

Now this skunk was active. His head moved from side to side and his feet were moving, but he couldn't spray. I started the truck with my right hand, put the transmission in drive, and away we went. Now you might be starting to feel sorry for that little skunk but regardless of the law, I was definitely going to let this fellow go. I love skunks and I had saved it, so to speak, by removing the can. I just wanted to have a little fun along the way.

There wasn't much traffic on the Mercer Circle Road, but when I got to the main highway a couple of cars saw us and passed giving us plenty of room. My first thought was to harass the office staff, but I remembered the DNR Ranger Station was empty and Carol, our secretary (who, as the sole female in the office, was often the victim of our jokes), was gone. I had this incredible opportunity, but how could I use it? Then I saw it, the Duck Shack Café. It was about noon and I ate there occasionally and knew all the waitresses. They were always joking and giving me a hard time. I pulled right up to the café's big front window and parked. By this time the skunk was quite lively, probably excited from his ride across town. As I got out with the skunk, tail in hand, I could see customers already pointing out the front window and when I got near the front door, I noticed a lot of movement inside and some shrieking. The waitresses and cook were already blocking the door. I went to the door and yelled, "Open up, I've got tomorrow's soup meat!" The waitresses were starting to scream and were not leaving the door, so I went to the front window and held up the skunk. Once the customers had also fully enjoyed the spectacle, I got back in the truck and drove away to release the skunk.

For reasons having nothing to do with the skunk, the Duck Shack Café has long since gone out of business, and of course today I would never consider such as stunt. But from that day forward until it closed several years later, the waitresses thought twice before they harassed me about anything.

The Elephant in the Room: Human Dimensions

"Unless someone like you cares a whole awful lot, nothing is going to get better, it's not."

—From *The Lorax* by Dr. Seuss

The Fox Whisperer

I first met Ursula Schramm back in 1980 at the North Country Folk Festival in Ironwood, Michigan. I had just gotten to the festival when I noticed a crowd of about 20 people in a circle on a hillside. Inside the circle was a woman in her late 60s standing with a sheep. She was stout and suntanned, with her long silvery hair pulled back in a bun. As I approached, I saw the woman grab the ewe, flip it up on its haunches, then reach down and pick up a pair of hand shears, not electric, and set to work shearing that sheep clean. She then picked up the wool, walked over to a spinning wheel and started demonstrating how to make yarn. First, she carded the wool with hand held-brush tools that combed the wool straight and even. Then she showed how the carded fibers were spun into yarn. In a rack displayed alongside her were some beautifully knit wool sweaters. I was impressed and curious, so I hung around a short time and asked her a few questions. I discovered I'd met an original back-to-the-land hippy from Iron County, Wisconsin.

One morning several years later my boss, John Olson, handed me a note detailing a call from a woman in Kimball Township, west of Hurley. She'd reported a red fox had been eating her ducks. As this was technically livestock depredation, we would have to respond. The problem was, she'd called from the public library where she volunteered, and she had no home phone. John said he didn't think she even had electricity, living so far back in the woods that the power company had no nearby lines.

This was the first fox problem I'd ever handled, but my youthful experiences in Iowa had prepared me for the challenge. To figure out where she lived I looked up her land in our go-to reference, the county plat book, a collection of maps that identify individual land parcels by owner. A quick look at the Town of Kimball section in the book, and I found her property, a 20-acre parcel off River Road. I went home and loaded up my personal fox traps and equipment.

I pulled up in the yard and saw a cottage with some vegetable gardens and a small orchard, several outbuildings, and a fenced area with sheep, lambs, ducks, and chickens intermixed. And there was Ursula.

Ursula said it had happened the morning before, when she was sitting in the outhouse answering the call of nature. The toilet door was broken, so she had an open view of her farmyard. She was sitting there watching her ducks, when all of a sudden, she felt an unprecedented but unmistakable urge to kill and eat one of her ducks. As she was registering this strange thought, the logical part of her brain saw a fox slip through the fence, run up to the very duck she was thinking about eating, grab it, and run into the woods. She said she thought that the fox had been lying there watching the ducks and its thoughts and desires had been picked up by her mind, a channeling or transfer of brain waves from the fox to her. Now this was a bit more supernatural than the wildlife stories I was used to hearing, but I was tactful and said, "Okay that's very interesting."

I then explained what I could do for her to prevent further depredation problems and that the DNR couldn't offer economic compensation for livestock losses from fox. I told her that foxes were too wary to be caught in cage traps. I added that I could set a foot-hold trap for her but she would have to keep her cats inside to prevent an accidental catch. She agreed to my offer of the foot-hold trap, saying she couldn't afford to lose any more ducks, as they were an important food source for her.

I thought about the situation. She was 35 miles away from the ranger station, had no phone and I was often out of the office doing fieldwork, so it would be hard for me to check her traps each day. Pondering on these factors, I asked her to help me by phoning early in the morning to let me know if the fox was captured. She said she had a car and could drive to a neighbor's house and use their telephone. We decided I would leave her a cage and a catchpole that she could use if I was delayed. A catchpole is a tool with a loop of adjustable cable at the end of a pipe with a locking handle. To use it she would slip the loop over the animal's head, and snug up the cable around its neck, securing the fox at the end of the pipe. This would allow her to hold the fox's head away so it couldn't bite while she released its foot from the trap jaws. Next, she could steer the fox into the cage, release the cable on the catchpole and shut the door. The catchpole I gave Ursula was from back home on the Iowa farm where we used it to catch young pigs. I felt better leaving that cage and catchpole because the weekend was coming up and I might be hard to get hold of. Of course, this was before cell phones were available.

I set two traps, making dirt hole sets, which are typically designed to resemble the caches my pet fox Zorro had made when he buried extra food. I carefully set a foot trap in front of each baited hole, staked them solidly, then sifted dirt over the trap, blending the dirt pattern and concealing it. I then gave Ursula my phone number and left.

The next morning Ursula hadn't called, so I drove to another ranger station to pick up a load of goose crates for a future banding project. When I got back John said "Your fox lady called and you caught the fox."

"Oh no," I said. "Has that fox been sitting out in the sun all day in that trap?"

"She told me no problem," John said. "Take your time, she handled everything like you told her to do."

As long as the animal wasn't in the sun too long, I really wasn't worried; my traps were short chained, well-swiveled and had offset jaws. When you took a fox out of one of them you could barely find a trap mark on its foot.

When I got to Ursula's, she said, "Come on in, sit down, and have a cup of tea." Her cottage was small with low ceilings, a bit cluttered, and was saturated with the aroma of fresh wool. Her spinning wheel stood in a corner. She said she'd built the cottage all by herself. As I sipped the tea I said, "So where's the fox?"

"Right there," she said and pointed. Then I saw the fox sitting in the cage below the piano, peering at us with a contented expression. She said, "It's a vixen." And went on to say that she'd had a long talk with the fox that morning and that the fox had said it would never step into another trap again. She had been playing German folk songs for it all morning and the fox really enjoyed them. And, oh yes, the fox said it promised her it would never eat any more of her ducks. I decided not to challenge her on any of this fox communication. After all, this situation was bizarre enough, so all I said was, "Okay, but just to be safe I think I'll release it down in Mercer, thirty-five miles away." I finished my tea, chatted a while longer then loaded the caged fox and left.

By the time I got to Mercer I had given quite a bit of thought to the fox and where I would release it. It was a nice looking animal, and while I knew a red fox had a home range of up to five square miles, I wasn't sure a transplanted fox would even stay anywhere near the place it was released. By next fall's trapping season, a prime fox would be worth over eighty dollars. I was a fur trapper myself and I thought, what would it hurt if I released it in a spot where I already trap fox each fall? What about Rodseth's field, an abandoned hayfield with county forest land on either side? I thought to myself that wouldn't mean I was necessarily going to catch it, as anybody could set traps near there. Further justifying my decision on release sites, I reasoned that the location would make a good home for a vixen. It was prime fox habitat with lots of forest edge and old fields with the related populations of small mammals a hungry fox would enjoy. Maybe she would stay there.

Months later on opening day of fox trapping season, there I was at Rodseth's field. I set three traps in my old favorite locations and was pleased to see fresh fox tracks around a mud puddle in the field lane.

I had high hopes, but the first morning check told it all. Every trap had been dug up, flipped over, and sprung off. Fox tracks were everywhere in the fresh dirt. All the bait was gone, so I carefully remade each set and re-baited. The second morning was a repeat of the first, and after a week I gave up and I pulled the traps.

I was saddened when Ursula passed away a few years ago. We had become good friends over the years, sharing stories over cups of tea in her kitchen. She told me that, as a 16-year old, she crossed through Nazi lines, alone, to escape her homeland during WWII. Her adventures eventually led her to America and her off-the-grid farm in Iron County. I respected her back-to-the-land lifestyle. Reflecting back, maybe Ursula and that fox really did have an understanding. At least I like to imagine that living so close to the land, Ursula developed a special connection to nature and wild things. I still wonder about that fox.

Ed's Amazon Python

One morning as I came into the office, Carol gave me a message John had jotted down on a notepad saying, "Jeff, you've got a dandy this time. A retired guy named Ed, from Ashland County, called reporting an Amazon python hanging around his place. He wants you to trap it."

I had to go that way anyway to set a bear trap at a farm corn damage site, so I worked the stop into my route for the day. When I got to Ed's place he was out in his garage. He asked me about the culvert trap on my trailer and when I told him it was for bears, he asked, "You trap bears?" and I explained that most of the complaints I responded to were about bear and beaver. "Oh, beaver too…you don't say," Ed noted. As we chatted, Ed explained he'd lived and farmed in the area all his life and that his wife had passed away a year earlier.

Ed told me he'd found tracks from the Amazon python on his dirt lane and in his gravel pit. I tried to explain to Ed that pythons can't live in Wisconsin, as our climate is too cold and they would die from exposure in the winter. I suggested he might have a track from a pine snake, which is our state's largest snake, and completely harmless. "Oh no, it's an Amazon python, all right," he responded; he could tell because the tracks were too big for a pine snake. I further explained to Ed that I thought it unlikely that a python would be able to travel the thousands of miles from its normal range to get to northern Wisconsin. Undaunted, Ed said, "Maybe it escaped from a traveling circus." I could see he wasn't giving up on this and then he looked me in the eye and said, "I have grandkids, how would you like it if that snake wrapped around one of my grandkids and squeezed the life out of him?" I suggested he show me the tracks. We walked down the hill into his gravel pit and he said, "Right there, that's them. They were plainer before the rain; you should have been here yesterday."

I didn't see much. There was a little ridge in the gravel and I thought it could have been anything. I asked, "Are you sure that's not a bicycle track instead of a python?"

There was a pause, and then he said. "I wasn't sure you'd come so I already set traps for it."

"Okay, Ed," I said, "Show me your traps." He led me to the other side of the gravel pit where I saw he'd set three foot-hold beaver traps on the ground. He'd secured the foot traps to concrete blocks with heavy wire and poured bacon grease on the trap pan triggers. He told me the snake would smell the bacon grease, come up to the trap and as it licked the bacon grease off the trap pan, it would snap the trap over its head and die. I thought things over for a moment, then said to Ed. "That should work fine, it's the same thing I was going to do. Let me know if you catch it." I then got in my truck and drove away, shaking my head.

A week later Ed called to report that the python had not come back but he had a bear that needed to be removed and some beaver down by the creek that he would like me to trap as well.

Northwoods Nessie

One day a man called from Long Lake, one of Iron County's more developed lakes. He was a year-round resident who owned one of the many lakefront homes. "I know you're going to think I'm crazy, but I had to call you," he said. "I saw something yesterday morning right out of my front window that's been bothering me ever since. When I saw it, I found my binoculars and watched it for at least five minutes. I swear I know what I saw and I know I'm not crazy, but you're going to think I am."

"Okay, what did you see?" I asked.

"Well, it was a long snake-like serpent thing, maybe 20 or 30 feet long. It swam like a snake, undulating just under the surface. It was black and thicker than a football. I didn't see any legs just the top of its back. It went right down the middle of the lake and the around the corner and out of sight."

Well, I knew this guy, and he wasn't crazy. "How about a log that's adrift in the wind?" I suggested, thinking maybe the log was waterlogged and bobbing up and down.

He thought not, saying it was definitely alive and moving. He went on excitedly, "It would turn. Besides, yesterday morning was calm. This thing could move and move fast." We talked a while longer, and then I hung up puzzled. I thought about his call all day . . . a long snake-like serpent . . . hmmm? Then it hit me, I had seen this "creature" myself years ago. Mystery solved!

I called him back and explained that he had seen a family of river otters traveling together, the mother in the lead and the young strung out behind her, nose to tail, in a straight line. The young ones would have been nearly full-grown at that time of year and they can have litters of up to five. Each otter would be about three feet long, so strung together, that would put your "serpent" at about 20 feet long and about as thick as a football. "Oh, yes," he said. "That's it, you just allowed me to get a good night's sleep."

Green Eggs and Ham

During the 1990s, wildlife supervisor Fred Strand out of Brule, and endangered resource biologist Sumner Matteson from Madison started an effort to enhance the population of common terns along Lake Superior. Ironically, common terns in Wisconsin are not at all common—they are listed as state endangered. After monitoring the population for a few years, Fred and Sumner decided to improve their nesting habitat to help recover the population. They selected a small artificial island near the Ashland city ore dock on Chequamegon Bay, which had a few common terns nesting on it. After contacting the owners, they got permission to remove the woody brush, and provided more gravel substrate and shored up the cribbing. This prevented erosion and made it into prime tern nesting habitat. I had helped Sumner with tern nest survey counts along Lake Superior and assisted with chick banding, which was done at their nests on the small island. However, the island had its share of problems. Some years we had great productivity and were hopeful the population was expanding. Other years, great horned owls or mink would target their gravel nest scrapes and kill and eat the vulnerable chicks.

We tried a number of measures to aid the terns' success. One year we tried to expand the colony to a second site and I helped Fred install common tern decoys and audio tapes in an effort to coax terns to nearby islands we thought provided additional nesting habitat. Some years, the predator problems were solved with trapping and fencing, and later the island was rebuilt with tighter fitting logs to make it more difficult for mink to access. While not expanding their numbers every year, the terns were holding their own.

About the time we had the island turned into prime tern nesting habitat, another competitor arrived on the scene, the ring-billed gull. The gulls arrived earlier in spring than the terns, and began to take over the terns' island, building their nests first and harassing the terns when they attempted to establish their own nests. The island was only about 1,000 square feet and about 20 ring-billed gulls were nesting on it. The gulls were about twice the size of the terns and were habitat generalists. The terns had nowhere else to nest.

Ring-billed gull populations were doing fine, with thousands nesting along Lake Superior with abundant nesting sites and good productivity. Fred and Sumner decided to take action and wrote for a special permit from the U.S. Fish and Wildlife Service to destroy all the ring-bill gull nests on the common tern island. My job was to visit the island every few days and destroy the gull nests. Like many birds, ring-billed gulls can re-nest once or twice if their first nest fails, but after a few tries they give up or move to a new site. Ashland County was my work area, so every few days I would drive to Ashland, launch a canoe and paddle out to the island, and break the gull eggs and scratch out the gravel nest scrape cups the gulls had made.

The eggs were buff-colored with brown-and-green mottling. The first time I broke some I marveled at how beautiful they were

with their bright orange yolks. They didn't have a lot of white in them but they looked fresh and were a bit smaller than a chicken egg. Now, my mom taught me not to waste food. Also, I had read in books that the French voyageurs, when traveling through this country during the heyday of fur trade, lived on a quart of dried corn or peas per day, occasionally supplementing their diet with a loon, duck, or gull egg when available. So, on my next trip out to the island I took along an empty egg carton. I filled it up then broke the remaining gull eggs.

The next morning at home in Mercer, I woke up first and started breakfast for myself and my ten-year-old son Zach. I got the skillet hot, fried some ham and then I cracked open and dropped in four gull eggs. Just then Zach walked into the kitchen. Half awake, he looked at me and asked, "What stinks?" Then it hit me too, the whole house smelled like rotten fish. I realized that the eggs would be inedible, fishy and probably laden with heavy metals to boot. We had to open the windows to air the house out and Zach happily settled for oatmeal that morning.

The good news was that Fred and Sumner's idea worked. That season the gulls finally gave up, and a good number of terns nested successfully. That island has since produced hundreds of common terns. Also, shortly thereafter, Zach took over the breakfast cooking and I was demoted to dish detail.

It's Only Three Tulips

In the 1980s we didn't have radios in our trucks. Fire control and conservation wardens did, but not wildlife management. If

I was on the road doing animal damage work, I tried to line up my stops to be more efficient. A couple of times I had driven all the way to Ashland or Mellen in northern Ashland County, then returned to Mercer only to find out someone had called late and I had an animal in a trap back in Ashland County and had to drive all the way back. John Olson solved this problem by getting me a phone credit card call number. At my last stop before heading back I would simply ask to use the phone, call the office, and ask Carol at the front desk if I had any messages. This was, of course, at a time when people actually answered agency phones; there were no answering machines at the DNR back then.

One day I called at the end of the day's run and Carol said "John's in and wants to talk to you." She transferred the call and John said, "I just got off the phone with Ida Rominski from Butternut and she about chewed my ear off. I could hold the phone a foot from my head and Carol could hear her in the next room. I'm sorry but I told her you'd stop by, you're still over that way aren't you?"

"Yes," I said, "What's her problem?"

"It's her flowers, the deer are eating them and I told her there's nothing we can do. They're right next to her front door, but here's the ridiculous part. It's only three tulips."

I jotted down the address and set out pondering how to approach this upset woman and came up with a plan. When I pulled into her yard, I jumped out of my truck and rushed to her front steps. She was already out the door and starting down the first porch step and that's when I saw the tulips. Her mouth opened in what I expected would be a dressing down, but I was faster.

Before she could utter a word, I cried out, "Oh my gosh, they are beautiful, what a shame, look at your tulips. My mother raises tulips too but yours are extraordinary, this is terrible. Look what the deer have done to them, I feel so sorry for you. What makes me feel even worse is that there's not really much I can do for you."

Barely stopping for a breath, I went on, "I could bring you a cannon gun which is an explosive gun powered by a propane tank that blasts off on a timed sequence. DNR uses them near Horicon Marsh to scare geese out of cornfields, but you'd have to run it at night when the deer feed and you'd never get a wink of sleep what with the blast going off every 20 minutes." I continued, "Just look at those beautiful tulips, what a shame. I'm always in such a bind, the hunters are mad at me, 'We need more deer,' they say. The farmers are mad at me. 'We have too many deer; look at the damage they are doing to my alfalfa field,' Nobody's ever happy. It's always the DNR's fault, and now, your tulips."

She was now on the last step standing next to me. I stopped my monologue, looking contrite, and finally gave her a chance to speak. She looked into my eyes and said, "Calm down young man, calm down. You're in such a frenzy, why don't you come in the house for some coffee and a piece of apple pie. It's only three tulips!"

Dangerous Dan

Dan, who was not dangerous, got his moniker on a canoe trip in the Yukon Territory, where, modifying a name from a Robert Service poem, we dubbed him "Dangerous Dan the Yukon Man." He was actually from Iowa and had grown up on the farm next to ours. My dad and mom sold a house to Dan's parents when our family bought a neighbor's 80-acre farm. We just wanted the land, so we sold the house and a few acres. I remember overhearing the discussion my parents had with the real estate agent. He came to our house and said, "I've got two offers, one from the Haskells, and one from a lady. The offers are close but the lady's offer is a little higher so I recommend you take it." I remember my mother saying we would take the lower offer. My dad objected, asking why we would do such a silly thing. She countered, "To get the two Haskell kids out of the trailer they live in and into a real home."

The Haskells were good neighbors. Dan's mom, Carol, drove a school bus and Dan's dad, Paul, was a mechanic in the road construction business. He was an excellent welder, and when Dad had a machine problem, he would go to see Paul. One time I remember the corn picker broke down in the middle of the night. A storm was coming and Dad was in a hurry to get the grain in the bins before the snow hit. He called Paul about midnight and in no time, Paul was there lying under the picker with his cutting torch. He cut out the damaged bearing, they put in a new one, and we were picking again. Dad would try to slip Paul a few bucks for his efforts, but he always resisted taking the money.

One day Dan called me and said he'd bought a new motorcycle. He hoped to ride the 400 miles to Wisconsin and do some walleye fishing with me. He had a four-day weekend open from his construction job, and I said, "Come on up, the fish are biting."

Dan arrived on a Friday night, and then the phone started ringing. I suddenly had several bear problems I would have to deal with, so I apologized to Dan—he'd ridden all that distance and I couldn't take him fishing. Dan asked if he could ride along because he wasn't keen about turning around and riding all the way back to Iowa right away. I said, "Sure, jump in," and all weekend we trapped bears in cornfields. I think we caught five with Dan helping me tranquilize, ear tag, and translocate them.

When Dan left on Monday, he said he'd had more fun than if we had gone fishing. He also said he'd decided to quit his job and go into wildlife conservation. I said, "Dan, you're crazy, my job is not all peaches and cream. With your union job you make twice as much money as I do; the state doesn't pay all that well. Besides, you're over 30 years old and without a college education. You can't get anywhere in my profession these days without at least a master's degree."

Dan said, "Yeah, I know, but the concrete work is destroying my back." It turned out he hated his job, so we dropped the subject. I remembered a quote my dad had found: "When you find the job you love, you never work another day of your life."

Against my advice, Dan quit his job and attended the University of Montana. He got his bachelor's degree in wildlife management and was hired by the U.S. Forest Service on a Canada lynx project in Montana, followed by another lynx project for the Colorado Division of Wildlife. He also worked on a loon project for the Alaska Division of Fish and Game.

One spring, a research colleague, Dr. Mike Meyer, needed some seasonal help on his common loon project. He needed to collect color band data on returning loons and later nesting success, capture, blood collection and banding. He asked me if I knew anyone qualified who might be available. I thought well, maybe I could get Dangerous Dan on the crew.

Dan was hired and worked three seasons before Mike picked him up as a graduate student. Dan and Mike studied the effects of shoreline restoration on wildlife and Dan graduated with a master's degree from Michigan Technical University two years later. He followed his dream, not my advice.

Bagging Marten

One fall, Dangerous Dan called me and said that his loon fieldwork with Mike was over for the season and Jim Woodford, from the Bureau of Endangered Resources, had hired him to go to Northern Minnesota to trap American marten. Did I want to come along? He needed a helper, Jim would pay, and all we needed to do was live-trap 13 females.

American marten were (and as of this writing) Wisconsin's only endangered mammal. Michigan and Minnesota had sustainable populations. In fact, both neighboring states had enough marten that their population model allowed a limited quota harvest trapping season for this valuable furbearer. Minnesota had agreed to give us 100 marten to translocate to Wisconsin to help our population recover but we had to travel to Minnesota to trap them. Jim's earlier crew had already caught and released most of the marten, but he wanted a 50:50 sex ratio. Males were much easier to catch than females, because females had much smaller home ranges, so when a trapline was laid out, chances were greater that the traps would be in male territories. Jim was short 13 females.

Our trapline started in Grand Marais, Minnesota and ran up the Gunflint Trail through the Superior National Forest along the Boundary Waters Wilderness Canoe Area. Jim had rented a small house for us in Grand Marais. He met us there and showed us the

cage traps, as well as the holding pens for feeding and watering the marten until their transfer to Wisconsin. Jim's last words were, "Don't come back until you have 13 females, and good luck."

It took us eleven days. American marten were not new to me. This project occurred after I had retired, but about 10 years earlier my last wildlife boss, Bruce Bacon, and I had been working with a science teacher at the Hurley High School named Diane O'Krongly. She would do about anything to get her science students out in the field and was always calling to ask us what we were doing and whether her students could get involved. Public education was a small part of our position descriptions, but we did all we could. We involved the students in our annual waterfowl nest box survey, where we visited each box and recorded its usage and replaced the nest material. They learned the difference between wood ducks, hooded mergansers, trees swallows and other wildlife box users, but Diane wanted a longer-term, more hands-on project to inspire her class.

We came up with a furbearer habitat research project at first focused on fisher and later shifting to American marten. The Iron County Forest, proposed as the project area, is over 170,000 acres, one of the largest county forests in the state. The county forestry department received five cents per acre from the state to do forest wildlife habitat projects. We generally used that money to build grassy wildlife openings, seed logging trails, plant oak trees, and conduct other wildlife habitat projects. The money could also be used for science-based outdoor education. Our initial idea was to catch a fisher, a member of the weasel family about the size of a large house cat, near each of the county's two school districts, Mercer and Hurley. We would involve the high school science students in the capture and radio collaring of fisher, as well as monitoring their movements using telemetry and tracking them in the field. The teams of students would develop a research hypothesis each year, do weekly field trips, write a report, and give a presentation on their findings. They would learn science, math, public speaking, and writing skills, as well as forestry management, tree identification, habitat typing, and orienteering using compasses and GPS. The program was well received by the county board and the communities. The kids did all the work and the project was sponsored through the North Lakeland Discovery Center, an environmental education center that served both communities.

After about five years of fisher study, Bruce found and documented a small population of marten in the middle of the county forest. He decided we should switch from fisher to marten, because their endangered status would mean the data on this population would be very useful to the DNR. I thought it was a bad idea— what if we don't catch any, after all they're rare—but I was wrong: within one week they had caught two marten and radio collared them. The project is still active today and as of 2016 they had captured 16 marten in nine years. The program is now sponsored by Iron County Land and Water Conservation Department in co-operation with the DNR, Great Lakes Indian Fish and Wildlife Commission, and Purdue University. The students have won two science achievement awards including a $70,000 Samsung Award for computer technology and some of the students have gone on to pursue careers in natural resources.

During all of this fisher and marten science research I'd stayed actively involved. The first coordinator of the fisher portion of the project was Terry Daulton, my girlfriend and later wife. She had a master's degree in environmental education and coordinated the trapping and telemetry work for the first four years. Terry's sister Diane worked for the Iron County conservation department and helped launch the program. The last year of fisher and following years of marten were coordinated by Zach Wilson, who has a degree in outdoor education and works for the Iron County...and, oh yes, he happens to be my son.

Between family and associates, I'd been part of many fisher and marten captures, had done or assisted with much of the drugging and telemetry work, and now here I was working in Northern Minnesota with Dan. Our trapline in Minnesota was in some beautiful country. We saw moose tracks every day and actually encountered two wolves as well as seeing numerous wolf tracks. We had good tracking snow and found good marten trapping locations.

The best habitat for capture locations was in old growth forests with a lot of coarse, woody debris on the ground. Marten are modest-sized members of the weasel family and are about as big as a fox squirrel. They feed mostly on mice and voles, but will take a chipmunk or red squirrel occasionally, preferring habitat with thick ground cover, not only for the mouse habitat, but as protective cover from predators such as fisher, great horned owls, and goshawks. Our trapline consisted of 30 live-catch cage traps dispersed over about 50 miles. We caught marten every day but mostly males; the females were few and far between, and, in the end, we caught 34 marten in order to get those 13 females.

One problem we encountered was determining the sex of the marten without taking a lot of time in our trapline check. Females are smaller, but it was hard to tell a yearling male from an adult female. In mid-September their fur was getting thick and longer as they began to "prime up" for winter. We'd hold the cage traps over our heads and try to look at their genitals from below to determine their sex, but it was very difficult to get a certain conclusion on the sex of the animal. We had a drug for tranquilizing, but it took a lot of our time to drug an animal and if we were wrong on a male, we'd have to use more time waiting for its recovery before releasing it and continuing down the trapline. Besides, a marten in a trap is pretty lively and it is quite a challenge to drug one. The best technique is to hook its tail with a stiff wire hook without getting bitten through the cage, then grab the tail with your hand and pull the tail outside the cage. This pulls the marten's hip against the cage. As you hold it there, your assistant injects the needle into its hip and pushes the plunger delivering the drug. Yes, it's as difficult as it sounds.

Then I had an idea—why not bag them without the drug? The next morning, I stopped at the local Gambles store in downtown Grand Marais and purchased a cloth mesh gym bag. It had a drawstring (like on a sleeping stuff sack cover) and was about the same size. We tested my new idea on the next marten we caught, slipping the bag over the end of the trap and tightened the drawstring. Dan then opened the trap door and the marten ran into the bag. When the marten hit the end of the bag, I grabbed the bag (with gloved hands) and squeezed it shut in the middle. Dan then released the drawstring, removing the bag from the end of the trap—we had our bagged marten. I lifted the bag and with both gloved hands slid my hand down the bag until the marten was tightly compressed at the bottom. Next, I clasped my thumb and index finger around its neck so it couldn't bite me. Dan then peeled back the bag, exposing its genitals so we could determine its sex.

From then on, we used this method. If it was a male, we released it with a gentle toss. I had to use the toss because if I didn't, they'd bite me every time. In fact, marten are so fast that one time I gave one a toss and it turned in midair, bit me, then let go and landed in a tree. If the marten was a female, I simply kept a gentle grip on the animal while Dan loaded a syringe with the drug and carefully injected the needle. We then took the standard measurements of the animal, put its radio collar on, and put it in a holding cage for transport to Wisconsin. We got so good at bagging marten that we did all of them that way, even the ones we knew were female.

One day we were gassing up at the local station when the owner asked us what we were doing. He had noted our Wisconsin government license plates when we'd stopped there every other day or so to get fuel. We told him, and he invited us out for supper on him. We met him that night at the Gunflint Tavern for hamburgers. He said he was on the board of the local public radio station and asked if he could send a reporter with us for the day with a tape recorder. He said she was a very nice young woman, which sounded good to us, so we checked with project manager, Jim. "Sure," he said, "but don't forget to mention how grateful we are to the Minnesota DNR for their efforts in allowing us this breeding stock." We had a great time with the reporter and even bagged a couple of marten for her; she thought we were crazy. After the show aired, everyone in town knew us.

Jim got his marten. Today, Wisconsin has an estimated population of around 1,000 American marten spread across the state. I like to think that our small project contributed to the larger efforts of the state, and federal agencies like Great Lakes Indian Fish and Wildlife Commission and U.S. Forest Service who are working to recover this endangered species in Wisconsin's northwoods.

The Legend of Rib Breaker

The story of Rib Breaker begins after the DNR launched a reintroduction program to restore extirpated trumpeter swans in Wisconsin. The last recorded sighting of these charismatic birds in Wisconsin was in 1937 when habitat loss and unregulated hunting led to their extirpation. The state's reintroduction program was led by my good friends from Endangered Resources, Randy Jurewicz and Sumner Matteson. They flew up to Alaska to collect swan eggs with philanthropist volunteers, Terry and Mary Kohler, traveling in style in the Kohlers' private jet.

Of course, Rib Breaker was just an egg back then, unnamed and anonymous, deceptively small and demure inside his smooth, white shell. The Kohlers helped the DNR collect 385 eggs between 1989 and 1997, making nine trips to Alaska's Minto River Flats and Nelchina Basin to find and collect eggs. These locations were selected because they were both under state ownership and had an excellent population of trumpeter swans, allowing Sumner and Randy to safely collect one or two eggs from each nest (swans normally lay five to seven eggs per clutch). After the field crew finished collecting, they brought the eggs back to Wisconsin for incubation. The project was a "reintroduction," which means that people actually move animals from one place to another. Reintroductions can be more controversial and unpopular with the public than "recovery programs," which are developed when an endangered species re-colonizes a previous territory on its own. But with the trumpeters we didn't have too much public opposition—after all who doesn't love a swan?

In the early years of the reintroduction project the eggs were incubated and hatched, and the young swans (cygnets) were reared in confinement pens. Care was taken to prevent the cygnets from seeing people to avoid them imprinting on humans and becoming habituated to people. If they remained "wild" their chances of surviving once released were greatly improved.

In "captive rearing" they were kept on their protected pond until they were nearly two years old. In their first autumn, they were health-checked, banded, and collared. They were wing-clipped to prevent flying away, and free to leave the pen. Many of them paired up and were kept together. We released several pair in Iron County wetlands. For some bird species, where a young fledgling learns to fly determines its future nesting location, its sense of home. Biologists call this site fidelity. Iron County had some great wetland areas with suitable swan habitat and the luck of the draw sent Rib Breaker our way.

All the trumpeter cygnets received U.S. Fish and Wildlife Service (USFWS) leg bands and easy-to-read numbered neck collars, allowing us to monitor the success of the program and follow the bird's migration routes. If a collared swan was sighted, the DNR was usually called, and the number and location recorded. During the last two years of the project we tried a decoy-rearing technique in Iron County. Instead of raising the cygnets in pens and later releasing them into suitable wetland habitat, the swans were raised right in the marsh where they would fledge. The Little Turtle Flowage, near Mercer, was one such site, chosen for its excellent swan food source, wild rice.

My friend Christa Reinert was hired to run the project at Little Turtle Flowage. She and her crew had constructed a small pen at the edge of the wetland that kept the cygnets safe from predators at night. Each morning, the crew would place an adult-swan-sized white plastic swan decoy in the water in front of the pen and open the gate. A pull line was attached to the decoy, which then led to an artificial, homemade, floating muskrat house about three feet tall. Christa or a member of the swan crew would be disguised inside the fake muskrat house, wearing rubber chest waders, snorkeling fins, and a harness attaching them to a floating inner tube inside the muskrat house. Their job 8 a.m. to 5 p.m. was to play swan vocalizations from a tape recorder while leading the flock through the wetland to feed on the lush vegetation. It was quite an ordeal for the crew. They had to keep track of their flock through a tiny peep hole in the hut while they roasted in the heat and endured mosquitoes all day. The point was to never let the swans see people. The cygnets were tricked into imprinting on the artificial decoy. At night, to the relief of the crew, the decoy was pulled into the pen and the gate closed to keep the young birds safe from predators.

I didn't get too involved with the project at Little Turtle until the cygnets started getting "gloved." Getting gloved is when a snapping turtle grabs a swan from under water by the foot. The turtle hangs on and eventually pulls the skin off the foot (like a glove) eventually killing the swan and eating it. Even if the swan escapes from the struggle, a gloved swan eventually dies. Three of Christa's cygnets were gloved within a week, the dismayed crew watching helplessly from the confines of their muskrat-hut costume.

You can imagine how upsetting this was to Christa, who had a lot of time and her professional future invested in this project. After the week of the gloving incident, Bruce Bacon and I got a visit. "You've got to do something about those snapping turtles," Christa pleaded. We came up with a plan. Snapping turtles were occasionally caught in the fishery crew's fish sampling nets, called fyke trap nets. We borrowed several from the fish crew and went to work setting them out in the Little Turtle Flowage waters. We started catching turtles, some quite large. We caught nine turtles in two days. When we thought we had their numbers reduced enough to satisfy Christa, we pulled the nets.

The swan crew continued the decoy-rearing project for two summers and reared over 40 birds. They were the most "wild" of all the birds the state had released.

My first encounter with Rib Breaker was not until after he was an adult, mated, and nesting on a beaver pond near Lake Ten, in a remote area in the Iron County Forest. At that time, Lisa Hartman was the DNR biologist in charge of the statewide swan fieldwork. Part of her job was to conduct aerial surveys each year by plane to document nesting. One morning, Bruce let me know that Lisa had found a nest from the air and wanted to go in on foot and count the eggs.

To get to the nest location, we had to canoe one mile up a narrow creek. When we got to the main beaver dam, the three of us crawled up and stood on the dam, easily spotting the huge swan nest in the middle of the pond on an old beaver lodge. The female was incubating the eggs and the male was standing guard next to her. Lisa looked them over carefully with her binoculars. She said, "The male has a crack in his neck collar. I wish we could catch him. I've got a new one with me I could put on." I said, "How do we catch him?" Without further explanation, she said, "If he attacks, grab him."

I thought she was joking, so I spread my arms in a gesture and called out, "Here I am. Come on and attack." I had no more than said those words when the swan let out a trumpet call and took off. He flew a half circle around the pond and then turned and leveled out, aiming right at us. I could immediately see he was zeroed in on me and before I could move, his 26-pound body hit me right in the chest at probably 20 miles per hour. I fell back, but somehow my instincts took hold and I grabbed him. I had no time to gloat that I had caught the swan, as he immediately took his wings and beat the holy heck out of me. It felt like someone had taken a wooden hockey stick, turned it on edge and whacked my arm five or six times. By the time we had him under control my arm was so swollen I could hardly move it. Lisa replaced the bright yellow numbered collar and we released him. Then she paddled over to the nest and counted the eggs. She called me the next day to see how my arm was doing and I played it cool, but it was sore for quite a few days. Although I didn't know it at the time, I had met Rib Breaker.

The next year Rib Breaker was back, this time nesting in a marsh along the Turtle River in a more populated area. The first

complaint came when he jumped into a moving motorboat and wing-whipped the passengers, biting one female passenger in the rump. She came into the Mercer Ranger Station to report the attack and said she had a black-and-blue mark where he had bitten her, even showing the bruise to Carol, our secretary.

After that incident, Mike Mossman, from the Research Bureau, came up from Madison and caught Rib Breaker, clipping his feathers so he couldn't fly until his next molt and moving him to a remote marsh for the rest of the summer. I asked Mike what technique he'd used to catch him? "I used your method," Mike said, with a laugh. "I let him attack me and grabbed him."

The third year Rib Breaker and his mate nested on Rice Lake, about one mile from Mercer. A few weeks later Gene Hickey, a resident of Rice Lake, called me to say a swan had him and his wife pinned down inside their house, and when they tried to go outside,

the swan chased them, biting and beating them with his wings.

I headed to the Hickeys' and immediately tried the "Here I am, come and attack me" technique, but Rib Breaker would not fall for it. I thought I should try something more dramatic, and I bent over, feigning fear and trying to create a wimpy posture. I would walk forward and then run backwards like I was afraid, and after several repetitions it worked. He came in for the attack and I grabbed him. Gene insisted on taking my picture holding the swan before I left. Bruce and I gave him another wing clip and released him on the Little Turtle Flowage where few people fished. A few days later, Gene Hickey stopped at the station and presented us with a framed picture of me holding the swan with a tongue-in-cheek caption about me saving the residents of Rice Lake.

After that we didn't have any swan problems for several years. We thought we were out of the woods on swan-human conflicts until one day when we got a call from Jeff Nehring, our local dentist. Jeff kept an airplane in a hangar at tiny Manitowish Waters airport, about 10 miles from Mercer. He told us that a large white swan was standing next to his white airplane and wouldn't let him get near it. He said that for days he had tried approaching his plane but the swan would always attack, and the one time he'd gotten past the swan and managed to take off, the swan followed and flew alongside the plane. Jeff was worried the swan would get caught in the plane's propeller and cause him to crash.

Jeff said he could see the swan standing guard by his airplane as we talked, so Bruce and I headed for the airport, a cage in tow. We wondered if this swan could have imprinted on the white airplane, thinking the plane was its mate and so was protecting and defending it. Could a swan fall in love with an airplane?

We knew that birds could imprint on inanimate objects. For example, the whooping crane migration project carried out by the International Crane Foundation in Baraboo, Wisconsin, trains their cranes to imprint and follow an ultra-light airplane from Wisconsin to the wintering grounds in Florida. If it worked on cranes maybe the same could happen to an ill-fated swan.

When Bruce and I got to the airport there sat the swan next to the white plane. We read the neck collar and realized it wasn't Rib Breaker. I walked up to the plane and I tried the wimpy retreat trick. It worked and again we were able to catch it and clip its wings.

Our last encounter with Rib Breaker occurred when a retired medical doctor in his 70s named Dr. Mosier was swimming near his home on Lake Barbara just north of Mercer. As he stroked along in the water, a large swan landed on the lake next to him, swam over to him and attacked. The good doctor barely made it to shore and suffered a broken rib and lacerations from the ordeal. It turned out it was Rib Breaker, and he was back at it again. This called for drastic action, so we called Madison, and Mike Mossman and Randy Jurewicz drove up. They caught Rib Breaker and took him away; his only option was to live his last years penned up as a captive breeder. This last incident of course was where he got his name.

In 1998, a woman named Pat Manthey took over supervising the swan fieldwork and oversaw the program for 16 years. She'd been to Alaska on one of the egg-collecting expeditions and continued to monitor the aerial flights to determine nesting success, conduct banding and give education talks. Once a year, swans molt their feathers and become flightless. If a pair has no young, both the male and female molt in July. If the pair had young, they stage their molt with the female molting earlier so only the male is flightless in July. This means that there is always an adult able to fly and defend the cygnets. July was when we held our annual swan round up. Pat would bring her swan crew and together with our Mercer staff and volunteers we would descend on the marshes to band and neck-collar swans. A DNR plane would fly overhead and direct us as we canoed through the heavy cattails herding the swans into a bay where we could corner and capture them with nets. It often turned into quite a rodeo with people leaping out of canoes and running soaking wet through the marsh with musky nets. After a few years we got pretty good at it. Each cygnet received a U.S. Fish and Wildlife leg band and numbered neck collar, and cracked or faded collars on adults were replaced.

Wisconsin assisted other nearby states in recovering their trumpeter swan populations by sending them birds. It was rare to have a troublesome swan, and most adapted to the wild. Rib Breaker was the worst.

Trumpeter swan recovery is a success story in Wisconsin. The trumpeter was taken off the endangered and threatened species list in 2009 and in 2016 the population estimate was over 4,700 birds in 25 counties. The sight of these beautiful birds is fairly common on northern Wisconsin lakes and wetlands, and there is nothing quite like the elegance of a white swan gliding on a calm lake at dusk.

Howling with Pam

It was just after daybreak and I was out checking a coyote trap near my cabin on the Flambeau. There had been a fresh snowfall overnight. The trap was just a quarter mile walk from the cabin, set at a natural narrowing of land frequently used by deer, fox, and coyote; a perfect funnel point for wildlife. We called this spot Four-Buck Pass for a legendary deer hunt during which it's said that a local gent bagged four whitetail bucks in one deer drive. As I approached Four-Buck, I noticed a trail of huge canid tracks: it was a wolf pack, and the trail headed straight towards my set! I rounded the bend and on the trail, not 50 feet away, I saw a wolf standing over the set. At first, I thought it was in the trap. I stood frozen in place. The wolf was facing away and as I watched it stooped down and rolled its neck and shoulder over the lure scent I had placed by the trap. Then it turned, looked directly at me and leaped into the woods. As I watched, five more wolves emerged out of the timber one at a time; each one stared me in the eyes, then dashed away. They were so close, I could see their expressions and I got the impression they were embarrassed, like a dog looks when it's been caught breaking a rule by its owner. It flashed through my mind that maybe they felt a little sheepish that I had snuck up on them. Each wolf was unique: the profile of its face, its coloration and size. The first wolf was dark with silvery tips on the guard hairs, an adult and probably one of the pack leaders. The next two were smaller, I assumed pups from that spring, followed by two larger animals that might have been yearlings. The last wolf to step out was really large and its fur was deep brown. When it saw me it laid its ears back in surprise. This all happened in a few seconds but to this day I can remember every detail because I was so astonished by the rare and humbling experience.

This was not my first encounter with wolves. The first wild live wolf I ever saw was north of Mercer in the early 1980s. I was driving my DNR truck and saw the wolf loping along the road towards me. I pulled over, and the wolf slowed to a trot, looked at me, and then turned and vanished into the forest. Since then, I've seen and handled many wolves and can remember every one of them. Wolf management was never a specific part of my job description, but I jumped at the chance to work with wolves whenever I could and I was lucky to be working for the DNR when wolf recovery efforts first began.

Wolves were extirpated from Wisconsin in the 1950s, and one of the last documented sets of tracks was seen by DNR biologist Bert Dahlberg in Iron County in 1962. It was a lone wolf and there was no evidence of an established pack. Wolves started recovering in Wisconsin in the 1970s when dispersers began moving into Wisconsin from Minnesota, and I was fortunate to get to see one of the first documented specimens, even though it was a victim of a car kill. It was an adult wolf that was hit in Douglas County, near the Minnesota border in 1975. I was heading for the office building at the Brule Ranger Station when I saw a large animal on the tailgate of Warden David Davidoski's truck. The whole DNR crew was around the truck gaping at it and when I got close I saw it was a wolf. I was immediately struck by the size of its long legs, large head and huge feet.

Our first wolf biologists were Dick Thiel and his assistant Ron Schultz. Dick had been fascinated with wolves since his college days and discovered the first re-established packs in the 1970s in Douglas County. Dick's DNR tenure spanned most of the 1980s and he has written two books about the recovery of wolves in

Wisconsin and his experiences studying them. Ron was a coyote trapper who was a wildlife major in college and he perfected the trapping methods and modified the equipment for capturing and radio collaring wolves. Our next wolf biologist was Adrian Wydeven, who worked through the 1990s and 2000s. During his tenure, wolves in the state reached both the federal and state wolf population recovery goals. Adrian has also published several chapters in a book about the recovery of Great Lakes wolves. Ron Schultz worked with both Dick and Adrian, and has logged hundreds of wolf captures to his credit, along with a long list of published and co-authored wolf research papers.

I heard my first wolf howls one night with Dick and Ron. We were deep in the Chequamegon National Forest in the Bootjack Lake Pack Territory. Some pups responded to Dick's howl (it's amazing how good some people can get at howling with a little practice) but the pack leaders would not answer. Still, it was a life first for me and I was thrilled!

I handled my first live wolf with Ron a year later when I was assisting him on his research trapline. It was a five-month-old pup that we caught on a forest road edge. We ear-tagged and radio collared the pup before releasing it, using a trick Ron knew to make sure that young wolves could grow into their collars. He lined the inside of the collar with foam rubber which would deteriorate over time, ensuring that the collar would stay on the wolf as it grew. Ron said that it didn't always work; sometimes other pack members would chew the foam off allowing the pup to throw its collar. This time it worked, and the wolf wore its collar into adulthood.

Wisconsin had a wolf recovery plan with a goal of maintaining a population of at least 80 wolves for three consecutive years, but in the early years, getting to 80 was not easy. The recovery plan demanded that we closely monitor the wolf population to determine if it was declining or increasing. This involved track surveys, capture and radio collaring, and documenting sightings of individual animals or packs. We already had an annual furbearer track survey, which we had been conducting for decades. During winter, we would drive designated road routes called transects and count and record tracks. Wolves were added to these surveys and when wolf tracks showed up the data was sent to the endangered resources wolf biologists.

We also looked for wolf scat and tracks while doing other field-work, and would get reports of wolf signs from foresters, loggers, hunters, and trappers that we would try to verify. Eventually, all this data helped us estimate and account for our wolf packs. The endangered resources crew used small fixed-wing aircraft to locate radioed wolves and count wolf packs to get exact pack sizes. In winter, the leaves were off the trees making it easier to spot the packs and count the animals outlined against the white snow. Pack sizes varied from two to ten. In 1980, we had five packs totaling 25 wolves; in 1985, we dropped to four packs and 15 wolves, and then in 1990, numbers increased to 12 packs totaling 40 wolves. For about a decade the packs barely held on and recovery was slow.

My involvement was mainly follow-up survey work but I did discover one new pack. It was in 1998, and I was conducting for-estry reconnaissance in Northern Bayfield County near Siskiwit Lake. I was just getting my equipment out to record some forestry stand measurements when I looked up to see two wolves trotting down the trail. As I watched, the first stopped, squatted, and uri-nated, and then the second did a raised-leg urination. I assumed this was a pair holding territory. I reported the incident to Adrian and it was the first he'd heard of wolves from that area—thus the Siskiwit Lake pack was discovered.

Disease and poaching greatly slowed down wolf recovery. Par-vovirus (a fatal canine disease) showed up in 1983. At that time, this disease was new to canids worldwide and through experimen-tation with captive wolves, pups were known to be especially sus-ceptible to death from exposure to it. Pups were lost in mid-sum-mer in four of five wolf packs then known to exist in Wisconsin, and the obvious suspect was parvovirus, though never proven. We also had outbreaks of canine distemper, mange, and other para-sites. There was really nothing we could do about disease except monitor and let it run its course, hoping the wolves would build up immunities and stand a better chance of survival.

Poaching was something we could do something about, and Wisconsin's recovery plan addressed poaching with an education initiative and increasing fines for poaching. Between 1979 and

1988, 45 percent of the known wolf deaths were caused by shooting. This was fueled by negative human attitudes towards wolves caused by a history of misinformation, conflicts between humans and wolves, and tradition. Negative feelings spread like wildfire where isolated incidents of wolves preying on livestock had ranchers and farmers up in arms. Wolves also occasionally kill domestic pets and hunting dogs, and this understandably upset pet owners. Wolves naturally prey on deer (average of 20 per year per wolf) inciting hunters to worry about their favorite recreational pastime. Knowing that deer hunters frequent wolf territories while in pursuit of deer, we knew that if wolves were going to recover in Wisconsin, we had to convince hunters, farmers, and the general public to accept and "live with wolves."

Our first task was to convince the public that the DNR didn't stock wolves in Wisconsin. We had stocked animals including fishers, marten, elk, trumpeter swans, and turkeys, but wolves had come back on their own, probably as animals dispersing from nearby populations in Minnesota. Convincing the public that wolves and deer can co-exist without significant negative population impacts on deer was, and continues to be challenging. Wisconsin has a substantial deer population, but the largest part of the herd lives in the southern part of the state, while wolves are found mostly in the north where deer numbers are more modest.

Hunters often blame wolves for declines in the deer herd. Scientifically, it is well documented that severe winters have a far greater impact. Black bear, bobcat, coyote, and of course wolves all prey on deer. Most often wolves kill old or vulnerable deer. However, I did investigate one incident of surplus killing by wolves. It was in 1997, and was reported to Bruce Bacon by local loggers. It occurred on a winter logging timber sale near Lake Six in Iron County which had concentrated over 100 deer, drawing them into the sale area to feed on the freshly cut tops of the newly downed trees. Although the feed was good, snow conditions were poor for deer, and perfect for wolves. While the deer floundered in the deep snow, the wolves could run on the crust. When we got to the logging site we counted over 14 dead deer, many of which were not eaten. This "surplus killing" by wolves is not common, but it does happen. Dogs will sometimes do the same thing if given a chance.

An effort was made by all DNR wildlife personnel to get the facts out to the public through talks and printed news releases. We knew wolves were controversial and there would be a "social carrying capacity" as well as a biological carrying capacity for this species. Society had to accept the concept of wolves on the landscape before we could expect their population to recover.

To help with these efforts, private wolf education groups formed focusing on Wisconsin, initially the Timber Wolf Alliance (TWA), and later the Timber Wolf Information Network (TWIN). TWA's headquarters was located at the Sigurd Olson Environmental Institute at Northland College, and for a few years, my future wife Terry Daulton oversaw the program. These groups gave hundreds of wolf education presentations and wrote numerous articles about the biology of wolves. From 1989 to 1998 shooting deaths of wolves in Wisconsin dropped to 19 percent mortality—a 42 percent decrease. The DNR knew that timber wolf management would always be controversial but it seemed public attitudes were changing.

In April of 1997, I was among a group of 25 colleagues who met at Park Falls to review the 1996–1997 wolf census. After the data was tallied, Adrian Wydeven popped open a bottle of champagne, passed around small plastic cups and gave a toast. This was the third consecutive winter Wisconsin had exceeded the state's recovery goal. The state was home to 150 wolves and we could now de-list the species from endangered to threatened status. This effort between educators and biologists had proved to be a winning combination.

I should note however that we still have occasional issues with poaching which seem to wax and wane with the deer herd and political winds. After I retired from my permanent job in 2000, I volunteered to help Bruce Bacon radio collar wolves. The DNR needed information for some Iron County packs, so Bruce and I ran a trapline for about 10 days. We caught and collared two wolves: the Pine Lake pack female and the Circle Lily pack male. Unfortunately, the Circle Lily male was illegally shot the winter after we collared him. We found this out because radio collars have a setting for mortality. Once a collar is immobile for a designated period of time (often 24 hours) the collars are set to send a specific mortality signal to allow the researcher to find and retrieve the animal's carcass and collar. In this instance, the mortality signal led to the dead wolf and a shotgun wound to the abdomen was identified as the cause of death. I examined his foot where the original capture trap had been attached. There wasn't even a mark. After the fact, I re-caught the Pine Lake female in a coyote trap two years later and released her unharmed. Her collar was still functioning.

As of this writing, Wisconsin has a winter population (before spring pups are born) of over 1000 wolves. We also have a growing, stable, white-tailed deer population. It was a long journey with the work of a lot of dedicated people, but after several decades of extirpation, it looks like Wisconsin's wolves are here to stay.

Collecting data, doing field surveys, capturing and marking animals, and monitoring populations is the most fun and perhaps easiest part of wildlife management. As the issue of wolves so clearly illustrates, the sticky part of the job is when human attitudes, emotions, and opinions get in the mix. Wisconsin wildlife belongs to all its citizens and future generations as well. I found that to do our job right, we had to recognize human dimensions of the work. We had to find ways to communicate scientific information that would be understandable and useful to citizens. Whether it is related to management of deer, fox, swans, or "pythons," people and their needs have to be considered in the mix of decision making.

A wolf encounter from 2007 remains my favorite, probably because I shared it with a group of amazing friends. Six of us had planned the trip of a lifetime, a 434-mile canoe trip down the Horton River in the Northwest Territory of Canada to the Arctic Ocean. John Olson was one of the paddlers, along with his wife Pam Troxell. Pam was a college friend of Terry's, and in the mid-1990s when Terry was still working with the Timber Wolf Alliance, the Sigurd Olson Environmental Institute hired Pam to coordinate the program.

Pam and John were both great paddlers, and had done a lot of wilderness canoe tripping throughout Canada. Pam had always wanted to do an Arctic river trip so they recruited four of us to join them. John had some connections through a former colleague, a Canadian biologist, and he asked if there was any scientific data we could collect for his agency on our trip. His friend gave us the GPS coordinates of some wolf dens the Canadian government had previously surveyed from the air and asked us to check them out for activity if we got a chance. We never did find an active den at these locations but one morning, about halfway through our month-long trip, we woke up to the sound of wolves howling. We were beyond the tree line so it was all open tundra and we could hear some adults and pups behind us hidden in a small valley. Directly in front of us about a quarter of a mile away was a light-colored lone wolf up on a bluff. Pam was an expert at wolf howling, having practiced during night howling surveys in Wisconsin. She gave a low howl. We could see the wolf sitting on its haunches, and then its head tipped up and returned her call with a throaty, beautiful howl that sent shivers up our spines. We all howled with the wolf for over 15 minutes until we got tired of it. The amazing thing was we could see the wolf and watch it as it howled with us and occasionally get a howl from the rest of the pups and other pack members behind us.

I treasure this memory in particular, as Pam died later that year after a valiant fight with cancer. I think of her and those Arctic wolves whenever I visit Northland College and see the life-size statue erected in honor of her work for wolves. It stands in front of the Sigurd Olson Environmental Institute, and shows Pam with a canoe paddle in one hand with her other hand cupped next to her mouth howling at that wolf along the Horton.

Lions and Tigers and Bears, Oh My!

"You don't have to sit outside in the dark. If however you want to look at the stars, you will find that darkness is necessary."

—Annie Dillard

My first exposure to Wisconsin's black bears came in the summer of 1974 when I worked for the Brule River State Forest between college semesters. We had a yearling black bear routinely scattering garbage from cans in the south campground. I helped the superintendent set a barrel trap and we caught the bear the next day. The trap was made out of two 55-gallon drums welded together with a sliding door at one end that closed when a bear entered and pulled on the hanging bait sack. I was amazed at how easy it was to catch that bear, similar to the raccoon trapping I had done back in Iowa. I was even more entranced by the intense shine of its deep back fur and its speed and agility when we released it in a remote forest glade.

Years later when I worked for the Bureau of Wildlife Management, I had the opportunity to accompany Bruce Kohn on his bear research trapline in Iron County. Bruce worked for the Bureau of Research out of the Rhinelander office, but each summer he set up a bear capture headquarters near Hurley. Bruce trapped for four years and caught 312 different bears. His crew had 30 barrel traps that they rotated at ten-day intervals over three traplines. It was on one of these trapline tours that Bruce's crew taught me how to foot-snare bears. If they came up against an elusive bear that evaded their barrel trap they would set an Aldrich foot snare. This was a spring activated cable snare that was buried and camouflaged along a trail, or in front of an enticing bait. When a bear stepped on the spring-activated release trigger a loop of cable flipped up and cinched around the bear's foot. The other end was secured to a stout tree to hold the bear. The art of setting one was to disguise it and use stepping sticks or rocks to guide the bear's foot perfectly to the center of the release trigger. I later met a Manitoba bear trapper at a research conference who taught me a lot more about foot snares. He replaced the black bear-sized springs with larger grizzly springs for faster cable throw, and added additional swivels and a car hood spring at the end of the cable. The hood spring would extend out as the bear pulled, cushioning the jerk when it hit the end of the cable. One challenge that had to be addressed when foot snaring was the selection of the tree the cable was attached to, as the bear was usually not too happy and frequently took out its frustration on the tree. I once had a bear doing damage in an apple orchard and it wouldn't go in a culvert trap. It kept flipping over the heavy cage trap, so I figured it must be a big bear. I told the orchard owner the last resort was a foot snare but the only anchoring location available was one of his valuable apple trees and it probably would have to be sacrificed. He said OK, but I don't think he believed me about the tree damage, or perhaps he didn't think I could catch the bear. The next day I had the bear and the anchor tree had no limbs left and four feet of its bark was scratched off.

Another challenge to snaring was removing the bear from the foot snare for processing or transport to another location. In the early years before I had a dart pistol, the only way to tranquilize a bear was with a pole syringe: a drug dosage syringe and needle mounted on a four-foot pipe. It was pretty easy if the bear was wound up around the tree with a short cable, but if it had a ten-foot circle and was lunging at you, things could get a little dicey. I solved this by having another person draw the bear's attention while I snuck into the circle to administer the injection. I eventually got pretty good at this but was very appreciative when I finally acquired a drug pistol and could fire a tranquilizer dart from a safe distance.

The University of Wisconsin–Stevens Point (UWSP) bear research crews capturing bears on the Apostle Islands found foot snares particularly useful. The islands were only accessible by boat, and once on the island, no motorized vehicles were allowed on trails, so heavy barrel and culvert traps couldn't be transported far from a dock. I trained several of the students in the art of bear snaring. I always said at the end of my bear trapping demonstrations that any bear that had four legs and could be coaxed to eat bait could be caught in a foot snare.

Bear Ecology 101

After we scraped the snow away from the den entrance with a small shovel, Jack said to me "I guess this one's all yours, Jeff, it's too deep and narrow for me to even get my shoulders in." We tied a rope around my leg and I wormed my way down the tunnel on my elbows. My head brushed the top of the tunnel and my shoulders rubbed along its sides. I was pushing the syringe pole ahead of me with one hand, following the beam of my headlamp. I could hear a low buzzing sound, a bit like the buzzing of a swarm of bees. I had heard the sound before; it was the sound bear cubs make when they are suckling. When I could see the sow, I lifted the syringe pole, aimed it at her hip and carefully inserted the needle and delivered the drug. I then kicked my roped leg up and down, the signal for Jack to pull me out.

I emerged into the winter cold, feeling it afresh after being in the warmth of the den and we sat down to wait the required seven minutes for the drug to take effect. When the clock had ticked down, I crawled in again, attached a second rope to the sow and gathered up the three cubs.

This routine was a regular protocol for me during several years when I occasionally assisted in bear research projects trying to tease out data on Wisconsin black bear reproduction and population dynamics. Den visits were only part of the work, but they were certainly a highlight.

Like people, bears are fiercely protective of their young and parent their offspring for longer than many mammals. Like human children, bear cubs play, roughhouse, and make a lot of mistakes when they are adolescents. However, unlike (most) people, bears only have sex during one time of year, and the males do not help bring up the kids—in fact they can be a significant threat to the cubs.

The adult female, or sow, breeds every other year, coming into estrus (fertility) between the middle of June and the middle of July. During that time, adult males are really on the prowl, seeking to breed as many females as possible. Bruce Kohn found that our Wisconsin females can breed at three-and-a-half years of age, unlike in some other states, where bears don't commonly breed until much older. Wisconsin also has a very prolific bear population. We know this because when a Wisconsin hunter registers a bear, a tooth is extracted from the carcass. Researchers cross-section that tooth and count the annual layers of dentine, which form rings somewhat like the annual rings in the trunk of a tree. In years when a female bear is suckling cubs, the ring is thinner because much of the available calcium is going into milk to feed the cubs.

These changes in annual tooth rings allow researchers like Bruce to learn a female bear's reproductive history as well as her age from a single tooth. The data from tooth aging, along with other factors, is critical to the Wisconsin DNR bear population model.

Another fascinating aspect of Kohn's work was his data on bear weights. During his study he caught 312 different bears 831 times and often recorded their weights. He found that they lost and gained pounds at a remarkable rate, and that individuals varied. He had one two-year-old male that only weighed 36 pounds, a real runt. Another male went from 140 pounds on June 3 to a whopping 290 on September 7, a gain of around 150 pounds in two months! During breeding season (mid-June to mid-July) some males clearly put romance above eating. One example was a manly bear who lost 91 pounds between May 22 and July 14. It wasn't just the males who could drop or add pounds, one sow weighing 263 pounds was recaptured a year later with three cubs and weighed in at 147. The caloric requirements of feeding three cubs was no doubt the main factor in this loss of 116 pounds.

Unlike (most) humans, black bears are polygamists, with females able to breed with several males each breeding season. DNA tests show a litter of cubs can have different fathers. After the breeding season is over, males and females go their separate ways and live solitary lives. An unusual adaptation in bears is that after the egg is fertilized, it goes into what is called delayed implantation for around six months. In the deep of winter (late January or early February) while the female is sleeping in the den, the egg(s) will implant and the fetus starts to grow. If the female is not healthy enough in winter—for example, having low body weight—the eggs may not implant and develop, but be absorbed by the female's body to protect her survival.

If the female is healthy, cubs develop and are born hairless and only weigh about eight ounces. The average litter size is three, but four is common and many cases of five have been documented. Over the years, I worked assisting Dr. Ray Anderson of the University of Wisconsin–Stevens Point and his graduate students who were studying black bear ecology, which took me into dozens of winter bear dens. We started the denning work in March when the cubs were around five pounds and fully furred. We needed to replace the collar batteries on the bears because they would only transmit for one year. The dens varied somewhat in size and configuration. Some were over six feet below the ground, a couple were high in tree cavities, and others were nothing more than shallow scrapes or depressions on the ground. One time we found a den right on the top of the ground in a shallow nest made from spruce branches. In the scraped-out area that formed this den were three healthy cubs. Bear dens are always neat and clean and the nest bowl is lined with grasses. Most of the bear dens were so inconspicuous we never would have found them without the radio collar's signal.

The collar technology included a waterproof transmitter, batteries, and an antenna that sends a signal to a handheld or vehicle-mounted receiver. The signal has a set frequency that becomes louder as a researcher gets closer to the animal. Our tracking method was to use the antenna, a map, and simple geometry to get a basic location for the collared animal by driving roads in the target area and stopping every half mile or so until a "beep" signal is picked up from the collar. Once the signal is steady, the antenna is pointed in the direction of the strongest and loudest signal beep, a compass bearing is taken, and a line is marked on a topographic map connecting the antenna location following the degree line of the compass. Then the researcher drives to a second location, perhaps a half mile away where a signal is detected, takes a bearing and plots a second line. The point where the two lines intersect gives the animal's approximate location. If more details on location are desired, or a den needs to be found, the researcher takes a hand-held antenna and follows the signal on foot. The "beep-beep" signal gets louder and, once the transmitter collar is very close, it thumps loudly, allowing the animal or its den location to be found. In theory it is easy, but there are complications. If an animal is underground the signal can be muffled. If there are hills blocking the signals or making them bounce, that can be a problem too. Sometimes the animal is in a location too far from a road for a signal to be picked up. In that case, budget allowing, airplanes can be called in to fly a grid over the suspected area. Even then, something can go amiss. Sometimes we would be standing right on top of the den with the bear directly below us and not know it. With persistence we usually were able to find our bear and then the exciting part of the job would begin.

I was often the one to crawl in to either administer the tranquilizer or take den measurements, ostensibly because of my smaller size at 5'8". If the bear was asleep it was easy, (that is if you think working your way into a snow-covered hole in the ground is easy). If they were awake it could get your adrenaline going. One time in the late 1970s, Jack Koch, the wildlife technician at Park Falls, darted a bear in its den and, unbeknownst to him, the dart failed to go off. After waiting for the drug to do its work, Jack started back down the den entrance only to be met by a very excited bear trying to come out! Jack scrambled backwards and, luckily, he had another dart handy. As the bear charged out of the den Jack made an excellent shot at the running bear and it soon wobbled to a halt. We re-collared that bear and carefully pushed and pulled him back into the den. Later, Jack surmised that the first dart failed because it was in his outside pocket and had frozen.

Another peculiarity of bears is that sows allow yearling young to winter in the den with them, and collaring all of them could be quite challenging, especially near the end of denning season if the weather got warm and the bears were active. Male yearlings can weigh 75 pounds with females topping out at around 60. When an adult sow with three yearlings were all jammed in the den together, we had our hands full getting them all tranquilized and sorted out. We would go for the sow first, then block off the entrance and take on the yearlings one at a time. A syringe pole and a flashlight were the best tools, and I was often the flashlight guy. The dens with newborn cubs were the most fun because the cubs didn't bite. We put them inside our parkas against our bodies to keep them warm. This was the only time cubs were fun to handle—they were like puppies.

The DNR learned a lot from the cub/den studies. Productivity and mortality rates were important to our black bear population model as we didn't want to harvest more bears during hunting season than were produced. We rarely found fewer yearlings in the den than cubs the year before, suggesting survival was usually good. I think this is because black bears are good mothers and are very protective. We also felt we had good habitat with plenty of food. Generally, that means our bears go into the dens with lots of stored fat, an essential for surviving the six-to-seven month winter fast, and for females, carrying the extra energy burden of producing milk for the cubs. I only remember one bear that starved during the winter. It was a female on Stockton Island in Ashland County, and the island's acorn crop had failed that year. The islands had mostly mature forest with poor forage for bears, which meant the Apostle Islands bears often had low body weights with little stored fat when entering their dens.

Bears have quite a reputation with the public as both a dangerous and a charismatic species, and when you work with bears you have to deal with these stereotypes. If I mentioned to people that I handled bears their eyes would widen and they would expect some hair-raising story. Quite contrary to this image, black bears are relatively easy to work with. Although they are rarely seen by the public, they can be captured with little trouble if you provide the right incentive (i.e. food). The black bear is a large, robust animal and, as such, it can wear a radio-telemetry collar with little risk to its health and survival. Acquiring data such as productivity is relatively easy. You simply follow their radio signals to their den and count them, as they conveniently go to sleep for the winter.

Though there have been a few dramatic instances where a bear caused my heart to race, I also spent many routine hours collecting data on bears and learning from my research colleagues who did the hard work studying their ecology. In the end, all the hours and analysis by university researchers and DNR field biologists helped Wisconsin maintain a sustainable population of bears and balance the needs of bears and humans. Of course, for me, it added an element of challenge and excitement to my career.

It's a Girl

When I started working with Ray Anderson's graduate students from UW–Stevens Point, the first one was Jack Massopust. Jack was placing radio-telemetry collars on bears, following their movements to study home range, habitat use, denning, and productivity. The DNR was particularly interested in his results on litter size, cub survival, and the age adults first start breeding—all data the DNR needed for our population model. Jack's study design required a 50-50 sex ratio. In late August, I got a call from him asking for help. He was worried he wouldn't have a chance to get his last two collars out—he was short two females. Jack had to return to classes in September, bear hunting season would be open in a couple of weeks, and bears would be starting to hibernate in mid-October. He had his 10 males but only eight females; he needed two more "girls" before hibernation started.

The wildlife tech at Park Falls and I had been helping him acquire his study bears when we were trapping and translocating crop damage bears. August is the month bears start preparing for hibernation through a process called hyperphagia. This is a fancy word for gorging on the foods at the end of summer to gain weight. Bears change their calorie intake from 8,000 calories per day to 20,000, allowing them to store fat to survive the winter. In Wisconsin, bears hibernate for six months or more, and during this time they live off their fat reserves. They don't urinate or defecate, and females give birth during this non-eating period. With over 30 percent butterfat in their milk, the cubs grow from eight ounces to five pounds before they leave their winter den. It's an incredible physiological process.

Wisconsin is a big dairy state and even in far northern Wisconsin counties like Iron and Ashland, farmers grow field corn, chop it in the fall, and then store it in their silos as winter feed for milk cows. The farmers like to harvest corn just when the kernels start to dry out and dent, right after the milk stage before the kernels have hardened.

Bears are omnivores, which means they eat both plant and animal matter. In general, they eat plants and incidental deer fawns in the spring, insects and berries in the summer and acorns and mushrooms in the fall. The fruits and nuts of shrubs and trees are called "mast" and are very important for their winter fat storage. Unfortunately for farmers, right about the time the bears go into hyperphagia, the corn is getting ready for harvest, and is at its sweetest "milk" stage. Bears aren't stupid—why dig around all day in the forest for acorns when you can walk into a 40-acre cornfield and gorge?

To set the stage, that August I was trapping bears out of cornfields to protect the farmers' crops when Jack's call for help came in. A couple of days later I caught a small bear, maybe 125 to 150 pounds, in a big culvert trap. Typical for bears at that time of year, this animal's fur was starting to prime up, getting longer and thicker for winter. This made it harder to see if it had testicles, which was about the only way to determine the sex of a bear that wasn't nursing cubs. I wasn't going to call Jack until I was sure it was a female. He was back at graduate school, and it was a three-hour drive to Mercer from Stevens Point.

I pulled into the Mercer ranger station a half an hour after quitting time and found all the staff had left for the day. I decided to tranquilize and ear tag the bear in the back parking lot, thinking that if it turned out to be a female, I'd just hose out the trap, feed and water it, and hold the bear in the trap overnight. That way Jack could come up and radio collar it the next day.

As I got my equipment ready, I noticed I was a little low on tranquilizer drug. I had a new shipment coming by mail but it hadn't arrived yet. I was using straight ketamine back then, 1cc per 20 pounds, so even a 100-pound bear needed 5cc. Smaller bears were the hardest to tranquilize as they could move around very quickly inside a big culvert trap, and avoid the syringe. The big bears usually filled the trap and had fur sticking out at several of the side holes, making it easy to hit muscle and deliver the required dose of drug. This was before we were supplied with drug dart guns. The only office with a dart pistol was Park Falls, because the veterinarian who supplied their ketamine felt sorry for them and bought them the gun with his own money. What I used back then was a homemade tool we called a jab stick—the pipe-mounted pole syringe I mentioned earlier. When alone, I would peer into a side hole in the culvert trap to line up with the bear. Then, holding a flashlight in my left hand and the pole syringe in my right, I'd aim the needle for the bear's rump and when I had the needle in the muscle I would push the plunger to inject the drug. This had to be done in a split second or the bear might jerk away, pulling free from the needle. Once I had the drug injected, I would record the amount of drug delivered, injection time, and when the bear first started to react to the drug. Additional data on how long the bear was "down and out," any unusual health conditions, ear tag numbers, and then finally the time when it had recovered (and was able to walk away) were all carefully documented.

I won't even describe what it was like with a bear in an Aldrich foot snare, spinning around on five feet of quarter-inch cable anchored to a tree. A testament to our training and care was that, of the hundreds of bears handled over the years, no bear was ever overdosed or injured.

This time, for all my good efforts, the first attempt with the pole syringe didn't go so well. I was sure I got some drug into the bear but I didn't know how much. I recorded the data and then waited seven minutes to see if the bear would react, watching for the typical signs such as rapid blinking, licking, and wobbling of its head. With a proper dose the bear would be down in seven minutes, so when the bear was still up after the allotted time, I decided I'd better give another dose. Again, bears aren't stupid— and once they get stuck in the rump with a one-inch needle by a guy holding a flashlight they wise up. On my next attempt, just as I started the syringe through the hole the bear spun around and bit the plastic syringe off with its teeth. That move cost me, as I had only 2cc of the drug left. I loaded up again with a new syringe and snuck around to the other side of the trap, deciding to omit the use of the flashlight, jabbed again and made what I thought was a perfect delivery.

I then recorded the data and waited seven more minutes. To my dismay, the bear was still up. It was blinking and licking, and wobbling its head back and forth, but it wasn't drugged enough to go down. It looked partially sedated but was still standing. When chemically immobilizing animals you have to monitor if the dosage is taking the animal going down or coming up. Going down means the chemical is spreading throughout its system and sending it toward the tranquilized state where it is immobile and you can handle it. Coming up means the drug is losing its effect and the bear's metabolism is burning off the drug, getting it back toward normality. Because I didn't know how much of the first dose had been delivered, I was not sure if this bear was on its way down or up.

The bear wasn't that far from the end of the trap where the small bait door was located. This door is about a foot square and made of welded mesh with steel hinges that open so that the trigger of the trap can be baited without crawling the entire length of the trap on all fours. Rumor had it that a game warden once set off a trap while baiting the trigger and had to sit in the trap all day before he was rescued. After that small bait doors were installed at the front of all culvert bear traps.

I opened this small door and used a stick to tap the bear lightly on the nose. It jerked but looked pretty sedated so I quickly reached in, grabbed an ear and attached a numbered ear tag. When the tag clicked shut, the bear jerked its head up and then moved back out of my reach. I thought, "Oh well, I'm out of drug anyway. I've done my job and this bear's not going to Jack Massopust because I can't determine its sex." I felt bad about not being able to help Jack, but under the circumstances I was lucky to have gotten one numbered ear tag on the bear. I decided it was probably a male anyway and I would release it and go home.

I got in my truck and drove a couple of miles north to a small field in the forest that a paper corporation used to de-bark aspen logs. Nothing grew at that field because it had a thick layer of ground up bark. I sometimes released bears in the field if it was far enough from their capture site, so I pulled in there, shut off the truck, and pulled the pin that allowed the trailer to tilt one end to the ground. Then I lifted the trap door. The bear had been leaning against the door, and with the tilted trailer, it rolled right out when the door came up. In its partially sedated state, it landed spread-eagle in the grass and I could clearly see it was a female.

Then I did one of the stupidest things I've ever done in my life. I thought, "I've got to get this bear back in the trap for Jack's research." Without further consideration, I reached into the back of my truck and picked up a 20-foot manila rope I kept for pulling drugged bears in and out of traps. It had a fixed loop on the end that tightened like a cowboy lasso. I flipped that rope right over the bear's head and in the next instant I realized my mistake.

My hasty plan had been to put this noose over the groggy bear's head, throw the end of the rope into the trap near the bait door and then run to the end and reach through the bait door, picking up the rope and pulling the bear back into the trap.

This might have worked if the bear had been more sedated, but as soon as the noose tightened on that bear's neck she reared up and began fighting the rope and me. She was definitely not drugged enough to be easily guided back into the trap. However, the drug's effect wasn't totally worn off, so I tried to pull her towards the door a couple of times, and almost got her into the trap. I succeeded in opening the bait door and got her close to the trap entrance, but she wouldn't enter or allow me to throw the end of the rope to the far end of the trap. The problem was the bear was pulling so hard I couldn't let go of the rope and run to the outside of the bait door to pick it up. She was excited and burning off the drugs rapidly and getting stronger by the minute.

In no time at all, it was the bear and me in the middle of the bark field, all alone. I was braced, feet planted at the end of the rope, she was running circles, and I rotated with her like she was a horse on a lunge line.

What was I going to do? I couldn't just let go of the rope, for fear she would run off and jump over a bank or get tangled in some brush and strangle herself. I alternated between cursing myself for not thinking the plan through, and wishing it was just a bad dream and I would soon wake up. But the reality was that there we were running in 40-foot circles; she wanted to be free, I wanted her free, but there was this rope problem we had to solve.

It wasn't long before the drug's effect was totally gone and her circles were getting faster. She was probably about my weight, but she was strong and getting stronger. After a few rounds of circling, seeming to tire a bit, she stopped and panted—before resuming her circular run. This became a pattern; she'd stop and rest then start running again. She was no doubt as terrified and exhausted as I was.

It was during one of her halts that I decided what I was going to do. It was a risky strategy, but it was all I could come up with. I had decided to get that rope off her no matter what the outcome. When she stopped to rest, I started working my way a little closer toward her, hand over hand, shortening the rope and reducing the circle. When I got down to about five feet, she stopped and her

body just froze up in cold fear. I bit my lip and starting inching down the last stretch of rope. We were staring into each other's eyes, both of us scared. When I got close, I slowly reached out to loosen the loop around her neck. I watched her eyes follow my hand. My index finger slipped into the loop but it wouldn't loosen. I was now touching her fur and body and I realized that I would have to release the tension on the rope a little to get my finger in far enough to free the noose. I could feel her fear as I pushed and curled my finger as far under the loop as I possibly could. When I was sure I had my finger in far enough to do the job, I reared back with all my might. As I pulled the rope back, she rose up standing in front of me and the loop eased and flipped over her head. She then turned, and in a flash she was across that bark pit and gone.

I just stood there holding the rope. I had this sick feeling in the bottom of my stomach. Then I walked over and sat down on the edge of the trailer for a minute and stared at that circle in the dirt where she had been running. It took a few minutes before my legs stopped shaking and I got in the cab and slowly drove back to the ranger station.

That fall was a busy one for us with a lot of bears in the corn and oats. We ended up trapping over 30 bears and the university eventually got their remaining two female study bears. Ray Anderson came up for one female and we collared it together. When we were done, we almost forgot to remove the magnet that activated the telemetry collar's signal. I didn't tell anyone about the lassoed bear. It wasn't a story to brag about. It wasn't a story I was proud of.

Superior Bears

"There is no better high than discovery."

—E.O. Wilson

Girl Scout Troop 237

In 1983 Jack Massopust's bear research project was winding down. Jack and Ray had documented some interesting things about Wisconsin's black bear population with their radio-telemetry work. Productivity was good with an average litter size of 2.4 cubs per litter, a high rate compared to other states in the Great Lakes region. Cub survival was excellent. If they found three newborn cubs in a den with a female, there were usually three yearlings with her the following winter. Home ranges were about 30 square miles for males and seven square miles for females, similar to what Bruce Kohn had found in his bear tag study in the 1970s. Another fascinating fact researchers found was that a sow would carve off a section of her seven-mile range and allow her female yearlings to live there but kick out the males to fend for themselves. A definite bias favored the females.

Jack monitored some of our translocated nuisance and damage complaint bears to see if they stayed put or returned to their home territories. Some headed right home, even within a few days, but he found that 64 percent stayed where they were released or moved into new territories. The age of transplanted bears made a difference, with older bears having greater site fidelity and returning home more often than younger bears. Bears that were moved farther than 50 miles had less chance of returning and, interestingly, the nuisance bears that did return were rarely a problem again. We hypothesized the negative experience of getting caught and handled by humans had a conditioning effect that caused the bears to fear humans and the capture sites.

One day Jack Massopust called me to ask if I'd help him catch and radio collar a female with two cubs out on Stockton Island, in the Apostle Islands National Lakeshore. He suggested I bring a smaller, lighter-weight barrel trap because we would have to transfer the trap by boat and carry it from the dock onto the island by hand.

The next morning Jack and I loaded our equipment into a National Park Service boat and headed for the island, a 12-mile boat ride. The park ranger on Stockton met us at the dock. His name was Brent McGuin. He looked to be in his 20s and was decked out in a gray uniform shirt with green pants and a regulation "Smokey Bear" hat. We looked rather scruffy standing next to him in our Carhartt pants, ball caps, and work boots. He said the bear was fairly docile, that she only had one cub with her, and he had no idea what had happened to the second. Brent said he saw her several times throughout the day and that she was becoming a nuisance around the garbage cans and campground. As he finished explaining the situation, a camper approached us and said he'd just seen the bear and cub—they were at his campsite.

The four of us each grabbed a corner of the barrel trap and hauled it up the trail to the campsite. The bears were standing about 30 yards away while we baited and set the trap. We quickly headed back up the trail, and in fewer than five minutes we heard a bang as the door slammed down. We returned to the campsite, saw the captured bear, and then ran the cub up a pine tree next to the trap. We processed the sow first, tranquilizing and collaring her, and recorded all the measurements Jack needed for his graduate project such as total length, chest circumference, paw, and skull size. Then we took a blood sample, tagged her ear, and weighed her on a hanging scale before we put her back in the trap to sleep off the drug.

That task done, I put on my climbing harness and tree spikes and flip-roped up to the cub. I tied in next to it on a limb and tranquilized it with my pole syringe. Jack and Brent held a tarp below me and caught the cub when it fell. It was a small cub, maybe 20 pounds or so. Jack had an expandable radio collar left over from a deer research project that Ray wanted him to try on the cub. It fit loosely and we both agreed it wouldn't last long (a day later Jack found it all chewed up). We released the bears and I headed home, satisfied that we had done our job. The National Park Service had its first radio collared Wisconsin bears.

About 10 days later my boss John Olson got a call from Bill Ferraro, Chief Ranger for the Apostle Islands. Bill told John that the collared sow's behavior had gotten progressively worse, and the university research crew was unavailable. While earlier the bear had showed tolerance around humans by occasionally walking down trails near campsites, now she was out of control. She had ripped open tents, destroyed coolers, even climbed trees and torn down suspended food packs. The park feared someone could get hurt, so they closed Stockton Island to all visitors and evacuated the campers at Anderson Bay and nearby Quarry Bay campgrounds. All the campers were gone except some Girl Scouts.

John said Bill sounded desperate, so we decided to head up there at 5 a.m. the next day and take along Charlie Zinsmaster, our seasonal employee. John said, "You'd better practice with that new drug pistol. We might need it." A few days earlier we had finally gotten a dart pistol. I'd built a carrying case for it out of a fishing tackle box. On the different shelves in the box, I'd added styrofoam pieces that were cut out to snugly hold the various components. The gun worked on a cylinder of compressed CO_2. When the trigger was pulled, the gas propelled a drug dart from the barrel. You could shoot several darts before the cylinder needed to be replaced. When the dart hit its target, a barbed tip entered the bear and the impact set off a .22-caliber blank cartridge in the tail of the dart. The explosion of the .22-caliber blank pushed the tranquilizing drug out of the dart and into the bear. The dart gun was brand new and I had never used one, so I took the pistol home and read the instructions carefully. The gun came with various sized darts for different drug doses, different needle lengths, plungers, and tails to attach for various applications. There was also a practice dart.

After I figured out how the gun worked, I set up a target on a hay bale in my yard and started practicing. I tried several distances, adjusted the sights a couple of times to fine-tune the accuracy and fired away. When I finally got everything adjusted to my satisfaction, I started concentrating on my last few shots. The gun's maximum range was about 30 yards. I noticed that the dart would drop somewhat at that range, but if I aimed about eight inches above the target I could compensate for the drop and hit the bullseye every time. At 10 to 15 yards adjusting to compensate for the drop wasn't needed. Shooting pistols wasn't new to me as I'd owned a .22-caliber pistol since my boyhood in Iowa. I shot the practice dart a few more times, then put in a brand-new gas cylinder to be all set for the next day.

Our plan was not to trap the bear. The Park Service said they had tried the barrel trap again but the bear was trap-shy from having been caught 10 days earlier. They said she would circle the trap and claw at the bait end, sometimes tripping the trigger by shaking the trap—one time rolling it over.

When we arrived at the Stockton Island dock, Brent was there with Jack's telemetry receiving box and hand-held antenna. That's when we officially met Girl Scout Troop 237 from Mount Pleasant, Iowa. They started filing out of the park's dockside equipment shed: 18 girls and four counselors. The girls looked to be about 12 to 15 years old, and the leaders maybe in their early 20s. They had arrived in Bayfield a few days earlier via chartered school bus and been ferried out to Stockton for a week of camping and hiking. The bus driver and troop leader had driven off and would not return until pick-up time days later. The girls were stuck, had no budget for motel fees, and consequently the Park Service had moved them to the relative safety of the equipment shed until a better plan could be figured out.

There were coolers and packs hanging from various trees, and in the shed, a tarp had been laid down and was covered with sleeping bags. The girls were quite excited. The bear had ripped open their tents and coolers, eaten much of their food, and ransacked their packs. They had seen this happen in broad daylight.

Chief Ranger Ferraro said he had to go. He'd be back with the boat at 4 p.m. to pick us up. That left John, Charlie, Ranger Brent, 22 Girl Scouts, and me. We got together and made a plan. Brent and the girls said they had not seen the cub for days and feared it was dead. Perhaps it had met the same fate as the first cub that disappeared.

Luckily, Brent had Jack's telemetry equipment, and he'd been plotting the bear's location twice daily to collect data for Jack so he could map its home range. Brent followed the standard protocols, tuning the receiver to the frequency of the bear's individual radio collar. If the bear was within range, say a couple of miles, he could pick up the signal. He would then take his compass reading and plot the direction of the signal on a map. Once he had his two or more readings for the bear's approximate location, that became another point in the home range data set.

The Girl Scouts had insisted the ranger stay in the shed with them at night and the signal beep from the bear's collar had been driving them crazy. Brent would leave the telemetry receiver on and when the bear made her rounds and came near the shed the beep grew louder, and when she went away it faded, no doubt acting on the girls' psyche like the music from *Jaws* where the tempo grew scarier as the shark approached. Every night the bear would circle the shed a time or two with the receiver beeping profusely, terrifying the girls, then it would leave and things would quiet down for a time. Nobody was getting much sleep.

The plan was that the Girl Scouts would stay at camp while the four of us headed out, following the telemetry signal. Brent told us that judging by the loudness of the signal beep the bear was currently about a half mile away. When fairly close, Brent and I would sneak in, leaving John and Charlie close behind. Brent felt we could get within 30 yards. This bear was not very afraid of humans, so we expected it would offer us a chance to dart it.

The drug pistol had not come with a holster, so I'd modified my .22-caliber pistol holster with a little padding to fit the gun. I loaded two darts with the required drug dosage, putting one in my pocket and one in the gun. I also stuck an empty syringe and extra bottle of drug in my other shirt pocket just in case.

Brent carried the telemetry equipment, John and Charlie carried packs, and I carried only the gun, and off we went. Right before we left the clearing and entered the woods I turned around and looked back. The Girl Scouts were standing there looking so hopeful, and that was when the pressure hit me. To a 14-year-old Girl Scout from Iowa this was probably the most dramatic event in her life. Spending all night with the bear outside and the receiver beeping, they probably thought they were going to die. The Park Service was counting on us, and John and Charlie wanted this to go well, and we only had until 4 p.m. I had to make a great shot and I started feeling really nervous.

We snuck through the woods, deer-hunting style, spread out in a line with me behind Brent. After about 45 minutes Brent whispered to me that we were getting pretty close. He'd turned the volume to its lowest setting and attached earphones so as not to alert the bear with the beeping sound. He stopped and whispered, "I think you should go on alone ahead of me from here, I think we are really close." I pulled out my pistol, bent low and crept ahead. We were in some dark woods, old growth pine or hemlock with brushy understory. I knew I only had one shot. I went one careful step, then another. Then I saw her; she was standing in front of a big pine about 30 yards away. I was standing next to a big tree, too, so I leaned against it to steady myself. I slipped off the gun's safety and brought the muzzle up. Suddenly, she was in my sights. I was about to pull the trigger when I remembered; at 30 yards, aim eight inches high to compensate for the drop at that range. I raised the pistol, steadied, and slowly squeezed the trigger. Bang! The gas exploded, sending the dart airborne. I saw it hit and heard the charge go off, injecting the drug. A perfect hit, right into the pine tree behind the bear. I had missed and the bear was gone in a flash.

The dart was embedded in the tree about eight inches above the bear's shoulder. I'd blown the shot and was devastated. John, Charlie, and Brent came running up to me. They'd heard me shoot. They hadn't seen the bear, but had been working their way in its direction and, hearing the dart fly by, they feared I'd shot at them. We tried to figure out what had gone wrong. The only thing we could come up with was that as I shot the practice dart the afternoon before, on my final sight adjustment, the CO_2 cartridge was getting low and probably had less power, causing the dart to drop more. Then when I installed the new gas cartridge peak power was restored and at 30 yards there really was no drop. I wondered if I'd just missed, but I liked the first theory better.

At this point I was waiting for John to say, "Give me the gun, you idiot, I'll try next," but he didn't. We made another plan. We'd try again and maybe the bear would let us approach for a second shot. It didn't work. We followed the bear for hours, but she was far too smart. She had figured out the barrel trap, avoiding recapture. Now she knew she had been shot at. She could smell and hear much better than we humans, so as we followed, she just stayed ahead…way ahead of us.

John called another planning session. Time was running out and we only had three hours before the boat returned to pick us up. The bear had been making big circles as we followed. Stockton Island is five miles long and one mile wide, covered in forest, so the bear had a lot of room to maneuver around us. The new plan was to leave me alone and have the rest of the crew keep following the bear with the telemetry receiver. I was just going to sit in the woods in a good spot and hope that if I was quiet and still while the others tramped around, the bear would come my way. It was a bit like a deer drive when hunters push deer to a good funnel spot, hoping a shooter hiding quietly will get a shot. We knew the chances of this working were slim, but we had no alternative.

They left and I waited 15 minutes, then started wandering around. I came to a spot where two game trails crossed on a ridge. I thought why not, the spot looked good. I backed off about 25 yards and sat down in some thick brush with my back against a tree. I raised my knees, resting the pistol on them so I could steady myself for a shot.

An hour went by, nothing—and then I heard something to my left. I slowly turned my head, saw black, and knew it was fur. I slipped off the safety and aimed for an open spot between two trees. Nothing happened, so I slowly turned my head again and then I saw her. She was looking back, no doubt toward the crew following her. She turned her head and started walking again. I started trembling like I had buck fever, the kind of nervousness kids often get on their first deer hunt. All the pressure from missing the first shot was causing me to come unglued. The bear stepped into my sights. I bit my lip to steady my nerves. I aimed for the center of her right hip and squeezed the trigger. Bang! The dart flew. Its tail was a white fluffy fleece material so I could see it had hit perfectly. The bear ran right past me with that dart bobbing back and forth imbedded in its hip fur. Then she was gone. I stood and started yelling but no one came.

I knew in seven minutes the bear would be down, but how far can a bear run in seven minutes? They can run up to 30 miles per hour on short spurts, so we needed to get to that bear fast. I kept shouting, still no one came. I shouted until I started getting hoarse. I knew the bear might only be sedated for 20 or 30 minutes before it woke up. That meant I needed to get to that bear within the next 27 minutes or it could get away. I started to search in the direction it had run. The forest was fairly open so I had good visibility. I was getting desperate, where were those guys? How far behind the bear were they? Did they choose this exact time to take a lunch break or something? I wasn't keeping time, but I knew my 27 minutes were ticking away.

I kept shouting and searching in the direction she had run. First, I heard her, then I saw her in a ravine. She had traveled maybe 200 yards before toppling over. I ran up to her and saw she could move, which meant she was coming out of the drug. I quickly reached into my pocket and pulled out my extra bottle of ketamine. I drew out a booster dose and injected her in the shoulder. She went down and I started shouting again, and finally I heard a reply.

Never was I so glad to see three guys in my life. Charlie quickly pulled out a hand ax and felled two small trees. We wrapped a canvas tarp around the poles, forming a gurney, and rolled the bear onto it. We all took a corner and lifted. We looked at Brent and asked where the heck are we and how do we get back to the dock? Thirty minutes later we all stepped out of the woods by the dock and 18 Girl Scouts came running, cheering, and surrounded us. I can't remember feeling more exuberance in my life. Of course, the birth of my son beat this euphoria, but this was a close second.

We put the bear in the barrel trap, and she was up in no time. The park boat appeared in the distance, and we were starting to

move our gear to the dock when two Girl Scouts, probably about 14 years old, approached me, holding out two potato chip bags. I looked at the bags. They were all chewed up, punctured with bear tooth marks. "Could we have your autograph?" they asked, handing me a Sharpie pen. I laughed and signed the bags. That was the one and only time in my life I've been asked for an autograph.

We hauled the bear to Manitou Island, tilted the trap off the stern of the Park Service boat, and down splashed the bear. She turned and swam to the shore of Manitou—her new home, we hoped. Jack Massopust later told me that the bear swam and island-hopped back to the mainland and was shot that fall by a hunter.

Ray Anderson had four other graduate students conduct bear research on the islands and I often helped them with winter bear denning work. They discovered that bear density on the islands was much higher than on the mainland, and that competition for food probably caused lower birthrates, and smaller body weights in both adults and cubs at birth.

Five years later Zach and I went home to the family farm in Iowa for Christmas. Mom insisted all her kids make it home for the holiday or come up with a really good excuse. My brother Brad showed up for Christmas dinner with a girlfriend he'd met at college. Sometime during the meal someone asked her, "So Merilee, where are you from?"

My ears pricked up when I heard her say, "I was born and raised near Mount Pleasant."

When I got a chance I asked, "So Merilee, were you ever in the Girl Scouts?"

She looked at me and said, "As a matter of fact, I was."

"Did your troop ever go on a camping trip to Stockton Island in Lake Superior?"

She said no, and I told her the bear story. Merilee said they had the same Girl Scout leader for many years and that the next time she was home she'd look her up. About two weeks later I got a letter from Merilee along with a photocopy of an article from the Mount Pleasant News recounting the tale of the Girl Scouts' adventures. Reading the article, you would have thought those girls were almost eaten alive by that bear. In the middle of the article was a picture of John, Charlie, Brent, and me, surrounded by the Girl Scouts, grinning ear to ear with that bear on the gurney.

The Bear That Rode the Ferry

This next story is really not mine. It's about a woman named Colleen and a bear with a scary disposition. In fact this bear was downright bad, and in hindsight we probably should have shot it, but we didn't.

Each year we sponsored a summer wildlife intern, usually from UWSP or UW–Madison, but once in a while from Northland College in Ashland, which specializes in environmental programs. In 1988, John Olson went to Northland's career day and interviewed several students for our internship and hired a woman named Colleen. In those days there weren't many women in the wildlife profession. At our state wildlife meetings and at the Wisconsin Chapter of the Wildlife Society meetings there was just a smattering of women in attendance. Many people still thought wildlife biology was a man's job, but Colleen didn't think so, and she was determined to prove herself. We soon found that no job was too tough for her. When we packed meat bags for the black bear population survey Colleen dove right in, rotten meat and all. One time, John tried to get some data sheets out of the truck but gave up when he saw there was heavy rain. A few minutes later, Colleen handed John the paperwork. John said, "Look at you, you're soaking wet."

Colleen looked at him and said, "Don't worry, John, it's only water."

The following year we had a lot of bear complaints. It was one of those years of poor natural food production for bears and we had a lot of corn, oat, and apiary (beehive) damage, as well as an above average number of nuisance complaints. That year Colleen was working as a Limited Term Employee (LTE), running a Youth Conservation Corps crew on the Turtle Flambeau Flowage doing nest habitat restoration for waterfowl, as well as helping me trap bears when I was over my 40-hour limit and needed a weekend off.

One Friday morning in August we got a call from the Ashland County Sheriff's Department telling us a bear had tried to enter a residence on Madeline Island in the middle of the night. It was at the residence of Dr. Donald Ramras, and the unusual and troubling part was that the Ramras family was inside the cabin cooking supper as the bear was attempting to tear the door off its hinges.

Colleen was scheduled to work that weekend, as I'd been on bear call the previous weekend and already had my 40 hours in by Friday. One reason we had hired Colleen was to save on our budget by eliminating my overtime, but when I heard this call I wanted to go in the worst way. However, Colleen was ready to go, having loaded a culvert trap on a trailer, bagged some bait and lures, and thrown in a couple of foot snares. I had no legitimate excuse to tag along. Colleen had trapped with us all last summer and half of this summer and probably had processed a dozen or more bears. Before she went off, I told her, "Good luck, call me when you get back and tell me what happens."

It took her about an hour and a half to get to Bayfield, a small town on the mainland shore of Lake Superior. There she joined the queue waiting for the two-mile ferry ride to Madeline Island. When she pulled into the ferry parking lot with a truck that said DNR in big letters on the door and an 8-foot-long culvert trap that said "Danger Bear Trap" on its side, people started checking her out.

Colleen was not your typical DNR employee of the time. She looked even younger than her 23 years, had coppery-red long curly hair and stood about five-foot four-inches tall. She wasn't exactly the Grizzly Adams profile the public stereotyped as a bear trapper.

After loading onto the ferry for the 20-minute crossing, she got out of the truck to stretch and stepped over to the boat rail. That's when people started approaching her with questions. She realized that rumors had spread like wildfire around the tiny community, and island residents felt they had a dangerous bear on the loose. It also occurred to Colleen that the men weren't exactly impressed that the DNR had sent her to rescue them. When she drove off the ferry onto the island things got more dramatic. The sheriff's deputy was waiting for her in his squad car as a personal escort to the incident site, and as the progression moved along led by the sheriff, the islanders following like a crowd going to a fire.

Colleen pulled up at the site of the attempted break-in, and a group of men approached her and immediately started telling her what to do. Colleen thanked the deputy and then found the landowner and his family. As the story stood, this bear really was trying to tear the door off the cabin with the people inside. What was unusual was there was no food source outside the house to attract the bear, just the family cooking supper inside. She explained to the family what she was going to do, then walked away and disappeared into the woods for a look around.

When she came out of the woods, the group of men approached her again. They told her they were experts because they were bear hunters and that she should set the trap in front of the cabin door so if the bear came back and tried to enter again, it would go in the trap and get caught. Colleen thanked them for their opinions and explained politely that she didn't agree with them and told them what she was going to do and why. The men listened but were not happy.

Colleen explained that during her reconnaissance she had found two day beds the bear had been using, a fresh trail with tracks and some fresh scat about 40 yards from the cabin. There was a small clearing next to the trail. She decided to set the trap there. She backed the trailer into the opening, tilted the trailer and then slid the culvert trap off. She stabilized the trap so it would not rock, explaining to the watching men that a wild bear survives on sight, hearing, and smell. When a bear sticks its head inside a culvert trap all three of these senses are greatly restricted by the trap, making the bear very uncomfortable, nervous, and cautious. A bear's greatest fear is humans, their only enemy in Wisconsin. Colleen pointed out that setting the trap in the clearing was a better location because it was away from human scent and activity. The chances of the bear entering the trap were much greater away from the cabin because it would be more relaxed and less cautious and the bear would also have more time to study the trap from the sanctuary of the forest while working up the courage to enter it.

She went to the bedding area and hung a two-pound mesh bag of Butterfinger candy bars on a tree limb about four feet off the ground (we purchase the unwrapped reject candy from the factory in 50-gallon containers). This was the freebie to get the bear's attention and get its taste buds working. She then added a second Butterfinger bag to a hook by the door and then the final bait inside at the front of the trap on the trigger. To finish off she raised the door, set the trap and scattered a few more candy bars between the door and the trigger. All the baits were lured with anise oil and liquid smoke, a smell bears love. She then cleared the area of all people and asked the landowner if he would check the trap in the morning and call her if a bear was captured.

Colleen got the call early the next morning, and when she called me, I told her I wanted to come along, saying, "After the story you told me last night, I've got to see this. I'll come as your assistant, as a non-paid volunteer. It's your show."

When Colleen drove onto the ferry with me in the passenger seat, we knew the cat was out of the bag. Even the ferry captain and crew knew about the bear capture. When we got to the trap the crowd was there ahead of us, and it was 20 questions again. They thought I was the DNR "bear man" that the men experts had been waiting for. I was ready for them too; my answers were easy: "I don't know, ask her."

We looked the bear over, estimating its weight at about 200 pounds, which is not huge but still a good-sized bear. It was all black, no white blaze on the throat or chest, and it was not shy. As we peered in from the front-welded grate end of the trap, the bear was about a foot from our faces and did not back off, unlike most captured bears that huddle at the dark end near the lowered door in total fear.

On the ride up we'd already decided to drug and ear tag the bear at the site. If the crowd wanted a dog-and-pony show, why not give them one? Besides, that would give the bear time to wear off the drug on the ride back to Mercer so it would be up and ready for release when we got there. Colleen brought out the drug kit, mixed up a cocktail of ketamine and xylazine in a syringe, and dispensed it into the barrel of a drug dart. She then brought out the drug pistol and asked everyone to clear the area. She didn't want someone hit by a ricochet. The crowd scattered and took cover. Colleen aimed carefully through a hole in the side of the trap. BANG—she made a perfect hit. I recorded dosage, delivery time, downtime, and ear-tag numbers as Colleen read them to me. She told the crowd, "Be real quiet, it should take about seven minutes."

As we were waiting, I walked around and talked to a few of the locals. Three women approached me saying, "Oh, you can't believe how happy we are that girl caught the bear. Those men were talking awful about her. They couldn't believe the DNR would send that young girl out to trap a dangerous bear. They said she'd never catch it. They're eating their words now."

After the bear was down, Colleen lifted the sliding door and locked it up. She then nudged the bear with her hand. When it didn't move, she attached a rope to a back leg and we pulled the bear out. The crowd moved in for a closer look. She carefully removed the barbed dart from its hip, cleaned the puncture location, then applied ointment to its eyes to prevent dryness. Next, she attached a numbered tag in each ear, then opened its mouth and checked its teeth. She spread its legs and pointed out to the crowd how to identify that it was a male.

We pulled the bear back into the trap by opening the small bait door in the front of the trap and pulling the rope through it. She closed both doors, loaded up, and we headed for home. We felt great, thinking that in addition to educating the public about our bear work, we might have scored a few points in the minds of the local men regarding women in the wildlife field. When we got to Mercer, we pulled into an area of county forest land near a small lake. It had been over two hours since the bear had been tranquilized and he was up and normal, the effects of the drug having worn off. Colleen pulled the cable raising the trap door and locked it up. We stood back about 10 feet behind the trap. Normally a wild bear shoots out of a culvert trap at full speed and never looks back, but this bear stopped at the edge of the trap and studied us for a moment. Slowly it climbed down off the trailer, stopped, turned, and looked us right in the eyes. It then turned and slowly walked into the woods. I felt very uneasy. I asked Colleen, "Do you believe in reincarnation?"

She said, "No."

"Well I don't either, but if I did, I'd say that bear could be a reincarnated Adolf Hitler. He just feels dangerous. I wonder if we should have shot him." The name stuck.

One afternoon that fall, John said, "Hitler's dead. I just registered him this morning. The local Methodist Church minister was bear hunting and Hitler walked into his bait beneath his treestand. The minister told me, before he shot, the bear just stared at him."

The Problem with Bears

"No matter how much you feed him, he keeps staring at the forest."

—Russian Proverb

The *Hunter* Bear

One hot summer day I returned to the office to find a phone message from Harold Schmude, the game warden from northern Iron County. The note said an elderly lady in Montréal, Wisconsin had a bear in her house. This seemed strange to me, but I called her and assured her I would soon be on my way.

Montréal is an old mining town, built as one of a string of towns stretching from Ironwood, Michigan some 25 miles west through the Penokee hills. As the towns grew, some, like Montréal and Hurley, developed along the highways into a kind of urban district. But Montréal kept a unique feel, proudly preserving its historic mining district with its quaint company homes and bungalows. It is not the kind of place bears normally hang out.

When I got there, I went to knock on the door, or rather what was left of the door. The screen door was hanging sideways by one hinge. When I gingerly called hello, the door was opened by a grandmotherly woman in her 70s. She introduced herself as Marge Barron, and also introduced me to her two housemates, a tabby cat and a grey feline with a white-tipped tail.

Marge started her story, telling me that last night, a hot summer night, she was sitting in her La-Z-Boy chair, watching her favorite TV show, *Hunter*. In the 1980s, this was a popular, weekly detective show where the private eye, Hunter, and his attractive female partner hunted down bad guys. Generally, the plot included Hunter getting beaten up or wounded but somehow, against the odds, they always won and foiled the criminals.

Marge explained that she fed her cats inside and they lived in the house as well—they were not stray or barn cats as was common in those years. That night, for the cats' further convenience, she had propped open the screen door a foot or so to let them come inside and eat. With them taken care of, she could relax and watch her show.

Marge recalled hearing some commotion by the door during the show, but she assumed it was the cats. At the same time the action was happening at the cat food dish, it was also happening with Hunter and the villains. Marge heard a bit more noise, but said that things quieted and she continued to enjoy the show. After a time, perhaps due to a commercial break, Marge said she suddenly got a strange feeling, like there was a presence in the room. She turned around and there was a bear, a really big bear, sitting on its haunches, gaze fixed on the television like he was also watching *Hunter*. Marge screamed. The bear leaped up in fright and dashed toward the screen door to escape. In its hurried flight, it ran right through the door and kept on going.

Now my job was to decide how to address Marge's bear report. Most bear problems are really people problems relating to food. In many cases, the DNR just gives advice about proper storage for garbage, bird feeders or food, and we rarely had to set a trap on nuisance complaints. I looked around the house and yard, noting that there wasn't a garbage problem or food attraction other than the cat dish. Bears have an excellent sense of smell (4 times better than a blood hound) so it must have smelled the cat food, followed its nose to the door, and wedged open the screen with its head before entering. And, like Marge, after eating it had decided to relax for a bit and watch Hunter.

Because the bear had actually entered Marge's house, this was a serious situation so I thought I should set a culvert trap. This bear's behavior was just too unusual. It was either really tame or just peculiar, and if it returned could pose a threat. Part of me doubted the bear would return—after all, this was a residential area and it could have been a fluke. Better safe than sorry, I thought. I had the trap ready to go in the truck, so I set it and headed home.

The next day Marge called and excitedly told me that the *Hunter* bear was in the trap. When I got there, I saw it was a big male, so I loaded him up and moved him about 30 miles away. Of the hundreds of bear complaints I handled during my career I only recall two other instances where a bear entered a residence. In both cases no one was home. Furthermore, none of the other bears I ever trapped had an interest in television. Maybe none of those bears had seen Hunter in action!

Don't Forget the Gun

One of my first bear cub incidents happened shortly after I transferred from Park Falls to Mercer in 1980. I got a call from Carl Marose, who lived in the town of Springstead in southern Iron County. Carl was a truck gardener who raised cabbage and cauliflower for commercial sale and I had helped him in the past when he had a deer damage problem in his cabbage crop. The deer would walk down the rows and take one bite from each cabbage, which meant he could only sell the cabbages for sauerkraut, a big loss of revenue for him. We solved his deer problem with propane scare guns that blasted a loud bang every few minutes, and with electric fencing.

This call from Carl was not about deer. He told me he had a dangerous bear that had him and his wife trapped as it circled their house. It even approached the front door. He reported that he had tried opening the door and shooting over the bear's head with his shotgun and it would run away, but then in 10 minutes it would be back again and he was afraid to go outside. He said he really didn't want to shoot it; he wanted me to shoot it, reinforcing this statement at the end of our conversation by reminding me to bring a gun.

As I pulled into his long, winding, driveway I immediately saw the cause of this odd bear behavior. There, at the top of an aspen tree next to Carl's front door, were two bear cubs. 10 feet from the tree was a tipped over garbage can. Thirty yards from the tree was Carl's garage, and attached to the garage, was a chain-link dog kennel with a black lab inside barking at full volume. Standing in the yard was a medium-size black bear sow that ran off into the woods when I drove up. "Carl," I hollered, "I'm sure glad you didn't shoot that bear. Come out here and take a look up there." Carl came out, gun in hand, glancing from side to side cautiously for the bear. Then he looked up and saw the two cubs.

My first step was to get Carl's gun unloaded. Then I got the dog out of the kennel and into the house basement so it would quit barking. We took up a position inside Carl's living room where we could see out but not be observed by the sow. We waited about 10 minutes and the sow bear appeared at the edge of the woods, looking towards the dog kennel. She rose up on two legs and sniffed the air, and then she slowly circled the house. Finally, she walked to the base of the tree, gave a whining sort of call, and the cubs started backing down the tree on command. Once something startled them and the cubs started back up, but down they came again. Once they were all on the ground they loped off into the woods.

"You know what happened don't you, Carl?" I asked. "That bear and her cubs were in the garbage last night and the

dog woke up and started barking. She got scared for her cubs' safety and sent them up the tree."

Climbing trees is a safety mechanism for frightened bears. Within a female black bear's territory, it's common for the sows to have designated trees they leave their cubs at from time to time. Researchers call them nurse trees—usually white pine or hemlock because they have rough, secure bark that cubs find easy to climb. I've found several over the years as they are easily recognizable by scratch marks and piles of cub scat around them.

I told Carl, "While your dog could see the cubs in the tree from the kennel, it kept barking, which kept them frightened and up in the tree. All you could see was the bear by the front door that wouldn't leave. That bear wasn't going to leave without her cubs." With a sigh of relief, I climbed back in my truck and backed out the driveway. Then I headed home and put my gun away.

The Baby Screams

Whether sensible or not, I have never feared bears. Lynn Rogers, a renowned black bear biologist in Northern Minnesota, once told me a black bear is just an overgrown raccoon. I had raccoons for pets when I was a kid, and I understood them, so I guess, like Lynn Rogers, I had a trusting attitude toward bears.

There have been very few people injured by bears in Wisconsin. One case was documented near Spooner when a Boy Scout was pulled out of his tent and received bites and scratches to his leg. In a second instance, a young girl was knocked out of her tree stand by a bear while she was deer hunting near Mellen. Another time, a pig farmer was injured in the leg and scratched up in an encounter. For the most part, bears will avoid conflict and contact with humans—they run away at the mere sight or smell of people.

In the late 1970s and early 1980s I conducted crop damage assessments to determine farmers' eligibility for compensation payments when bears got into their fields. I also walked a lot of oat and corn fields looking for trap placement locations, so I encountered a few bears face to face in those days but never had a problem. One time in a cornfield near the town of Butternut, I caught a 35-pound cub in a foot snare. When I walked up to the cub it screamed like a baby and the sow came running, but she just circled me in the cornfield. I only got a few glimpses of her now and then through the rows and I tranquilized the cub, ear tagged it, and then released it without incident.

One summer, John Olson and I handled five litters of cubs. It was a poor year for natural forest foods as late spring frosts had damaged the acorn and berry blossoms. We caught 88 bears that summer, and cubs were always a challenge. Occasionally the sow was in a culvert trap with one or two cubs but never all of them. The uncaught cubs were always lying down outside next to the trap.

John and I were good tree climbers and used the same equipment for retrieving bear cubs as we did for osprey banding and nest platform work. Other wildlife techs usually set cub traps, but they often didn't work as well. They'd have to leave the sow in the trap and hope they caught the cubs the next night, which was always an uncertainty. The landowners always checked the traps and called us if a catch was made, and if they told us there were cubs, we came prepared. I usually needed helpers—often the fire control guys came along if it wasn't fire season, and sometimes the landowner would help as well. If there were cubs, I would quietly sneak up on that trap, and when I got near, I would come running and yelling at a dead run, clapping my hands. This surprise approach was part of the plan to try and get all the cubs up the same tree, near the trap, so I only had to climb one tree. It usually worked, but not always.

One time in Ashland County, I had the sow in the trap and three loose cubs. John Bernier from DNR fire control had come along to help. We went in running and yelling, and as luck would have it the three cubs went up three different trees. The first one was in a cherry tree about 40 feet tall so I put on my gear and climbed up the tree. I sat alongside him and strapped myself on to the trunk with my safety harness. The cub weighed about 30 pounds and was out on a limb. I followed my usual procedures and had John attach a syringe pole and lowering rope to my main line and I pulled them up. The cub was about five feet away and he was snarling at me a little. I loaded the syringe with the proper dose, reached out, and gave him a shot in the hip. He jerked a little then settled down and I waited, plotting an angle to make a grab for the cub once he was sedated. John and the landowner were holding a small tarp off the ground below in case something went wrong and the cub fell. A few minutes went by, and then the cub started wobbling his head and licking his lips with his tongue, a sure sign the drug was taking effect. A few minutes later I reached out and patted him on the leg to test his response. He jerked a little, then let me touch him, so I slipped the noose of my small lowering rope over his back leg and tightened it. This way, I had hold of him if he toppled over and fell. Then I waited a bit more, and when I thought the time was right and he would be thoroughly sedated, I swung him out off the limb and into my lap. Much to my dismay, the timing was not right.

John Bernier loved telling the rest of the story, and I've heard him tell it many times. He always said, "Jeff, you screamed like a baby. You could have been heard a mile away." In my opinion, he was exaggerating a bit, but I will admit I was screaming because that cub was biting and scratching the heck out of me. I should have had leather gloves on, although those sharp canine teeth would probably have gone through them anyway. The cub bit my hands five times as I tried to hold on to him, and the last time I was looking into his eyes when he clamped down and started shaking his head to cause more damage kind of like a dog would do trying to "kill" a chew toy. At that point, I had to drop him. They caught him in the tarp, which broke his fall, but he leaped out and was gone.

The worst part was that after I climbed down and got my hands bandaged with supplies from the truck first aid kit, I had to climb the other two trees, drug those cubs, and lower them down. I put on some leather gloves and all went as planned. Then we drove to the emergency room at the hospital in Ashland. They asked if my tetanus shot was up to date and bandaged me the best they could, telling me they were just puncture wounds and I didn't need stitches. I still have those scars on my hand.

The Cub Who Remembered Me

I am convinced that bears can recognize human faces, and this story is my proof. John Olson and I had a sow bear and two cubs in a culvert trap over at the Kretzschmar dairy farm in Ashland County. The third and last cub was in a tree near the trap, so I climbed up to the cub and crawled out on the limb next to it. We were only three feet apart, and it looked me right in the eye, hissed, and snapped its teeth at me. I stuck the syringe pole in its rump to drug it, and a few minutes later it toppled out of the tree. John and the farm crew caught it in a tarp when it fell and we ear tagged this cub first. Then we processed the sow and other two cubs and put them all in what we called the circus trap, which is a cage trap made with gravel crusher screen that gave it a look like jail bars, allowing easy visibility for the bear and spectator. The standard culvert traps were dark inside, and gave the captive animal seclusion, but we had to re-set and leave the culvert in case more bears were in the area feeding. We headed back towards Mercer and by the time we got to the ranger station the bears were up and awake, looking uneasily out through the bars of the trap.

It was a Saturday and John said, "These bears are too cute. The kids are going to want to see these. I'm going home to get Dara and Devin before we release them. Why don't you get Zach?" When I got back with Zach a few more people had arrived to see the bears. Zach and the other kids ran up and looked into the cage while the bears were looking out at the people. I approached, and when I was close enough that the bears could see me, one cub gave a human like scream and put its two paws over its eyes. When I went around the corner out of sight the

cub uncovered its eyes and relaxed. When I returned a little later and the cub saw me it screamed again and covered its eyes. Someone turned to me and said, "It's you, Jeff. The cub only screams and covers its eyes when it sees you. Let's experiment." We did it a few more times and then John asked, "What's the cub's tag number?" We checked and sure enough it was the cub that had been in the tree with me. It appeared that it recognized me as the evil guy who had poked it with a needle in the tree. I have wondered if all its life it had nightmares about me.

Saved by Bandit

One Monday morning, John Olson got a call from Kent Goeckermann, the superintendent of Copper Falls State Park near Mellen, Wisconsin. Kent was also on the Mellen volunteer ambulance squad and the day before they had gotten a call from the wife of a local farmer named Ed Tafelski. Ed had been attacked by a black bear and Kent called us wondering why the Mercer DNR had not been contacted. Kent said the attack was the talk of the town and that the Duluth News Tribune had heard about it and sent a reporter by helicopter to interview Ed. The Ashland Daily Press had been there as well.

The nearest warden on duty that weekend was George Phillips from Drummond in neighboring Bayfield County. He had investigated, written an incident report, and had been interviewed by the Duluth paper. It had happened on the weekend, and apparently none of the authorities had realized that Mercer handled Ashland County bear incidents. Ed had been rushed by ambulance to the Memorial Medical Center in Ashland, where he was treated for a scratched leg, cracked rib, and a swollen and dislocated knee.

John and I set out right away, and we found Ed at home on crutches and with a cast on his leg. We sat down with Ed and his wife to hear the story of the attack. Ed had about 15 young pigs in a feed lot pen about 50 yards from his house. One side of the L-shaped pen was attached to a shed where he kept the pig feed. He fed his pigs in the morning and evening, and the morning of the attack Ed and his dog Bandit left the house together to feed the pigs. Bandit was about 40 pounds, a medium-sized brown-and-black mixed breed, probably half German shepherd. Uncharacteristically, Bandit immediately started barking, so Ed, fearing Bandit would frighten the pigs, tied the dog to a bent spike he had pounded into a basswood tree in the yard. Ed then mixed the pig feed in a bucket and entered the pen. When he rounded the corner of the L-shaped pen he came face to face with the bear. The bear charged, knocked Ed down and jumped on him. Bandit saw all of this and went wild jerking on the rope. In just a few moments, the jerking straightened the spike and Bandit ran across the pen and jumped on the bear, biting it on the neck and distracting the bear from its attack on Ed. The bear started fighting Bandit, swatting at him and knocking him around, but Bandit kept coming back, running, and biting while Ed crawled away to safety. Eventually, the bear ran into the woods. Ed felt Bandit had saved him.

We set one culvert trap on the wooded side of the pig pen and another on a forest trail we thought the bear might use. We also set cable foot snares further up the trail, covering about all the possible bear travel locations on the property we could find.

The next day we had no action in the traps, but the newspapers complicated our plans. Articles came out in the Duluth and Ashland papers with more details of the attack, including quotes from Conservation Warden Phillips. Regrettably, some of Warden Phillips' statements were not accurate, for instance, that bears were related to pigs so it was natural they would eat the same food as pigs, and that this bear was probably uncatchable. The statement that bears and pigs were related was total baloney, but what really got my dander up was his suggestion that the DNR probably couldn't catch this bear. Later that day we got a call from Margaret Levra at the Ironwood Daily Globe. She was clearly irritated that we hadn't given her a scoop on the story, and she wanted an interview ASAP. She had done several newspaper stories with John and me in the past, so we promised her that if and when the bear was caught, we would call her first.

Day two came and went, and still no bear in the traps—but another newspaper article came out, this time an interview of local citizens on the streets of Hurley, asking for opinions on what the DNR should do with the bear if they caught it. They interviewed five people and printed their pictures and answers. Two said it should be killed, two suggested relocation, and the last response was "depending on the temperament of the animal it should be put in a zoo."

Day three came, and I was starting to worry that maybe the warden was right after all, and we couldn't catch this bear. Then the phone rang, and it was Ed reporting the bear was caught in one of the foot snares in the woods.

We immediately called Margaret and left for the farm, meeting her there an hour or so later. The bear was a male, probably 225 to 250 pounds. It looked and acted like any ordinary wild bear. It had no deformities or injuries and wasn't thin or starving. We could not detect any physical reason it would attack a human. While bears can be dangerous, with 60 people killed by black bears since 1900 in North America, there were (and still are) no records of human mortality due to bear attack in Wisconsin to date. The nearest bear-caused death had happened in upper Michigan in 1952 when a three-year-old child was killed. Comparatively speaking, for every human death caused by black bears, 25 people are killed by snake bites, 180 by bee stings, and 160 by tornadoes. During my presentations to citizen groups on bear ecology I was always asked, "How dangerous are black bears? What should I do if I'm walking in the woods and encounter a bear?" I always ended my response with the fact that for every human death caused by black bears, 160,000 people are killed in traffic accidents, suggesting to the audience with a bit of tongue-in-cheek that if they came to the talk by automobile, they should be careful on the way home. I would reinforce this message by proposing that they enjoy a walk in our bear-infested northwoods tomorrow as it was one of the safest activities they could do...but I digress.

John and I pistol-darted the bear, removed it from the foot snare and loaded it into a cage trap and returned with it to Mercer. Margaret's headline the next day read, "Tafelski says, 'That's the bear.'" with a picture of Ed in his leg cast next to the bear.

John and I had to decide what to do with this bear. One problem we foresaw was that if we let it go free, we were likely signing the death warrant on at least 10 innocent bears. Public fear from this incident might mean that the next few bears to get into someone's garbage can or bird feeder would get shot by a landowner assuming it was the "Tafelski attack bear." We were also worried about the public relations fallout when people heard the DNR was releasing a "dangerous" bear. With all the publicity from this incident, someone was going to ask where we'd released it and we'd have to tell them, as it was public information. No one would want this bear released in their neck of the woods and we could also face a liability problem if the bear had another incident. We had just about decided our only choice was to shoot the bear when John had a great idea.

He'd read about a research project on bear hormones at the University of Illinois–Urbana. The project's lead investigator was Dr. Ralph Nelson, and he worked at the university's Carl Foundation, a medical research unit. It turned out Nelson was trying to extract the bear hormone that causes bears to go into hibernation to test for use in humans. Unlike other hibernators, bears don't lower their heart rate and body temperature much below human standards. Dr. Nelson hoped that humans with kidney problems could take the bear hormone instead of going on dialysis machines while awaiting kidney transplants. Dr. Nelson also had a grant from NASA to investigate use of the bear hormone in space travel, testing the potential to send astronauts for weeks or months through space in a hibernation state. This could eliminate the food, waste, and perhaps some of the mental challenges of longer space journeys.

John made the contacts, and yes, Dr. Nelson wanted our bear, but there was a problem. His staff said he was in the field studying polar bears in Churchill, Manitoba, probably best known as the polar bear capital of the world. He was far north on the shores of Hudson Bay with only a satellite phone and would not be back for two weeks. His staff relayed his request that we keep the bear until he returned in two weeks.

Margaret Levra wrote another article, this one about Dr. Nelson and his research project. Our bureau supervisors in Madison loved the plan, but we were left with the problem of where and how to keep this bear for two weeks. In the end, we decided I'd take care of him at my home on the outskirts of Mercer. At the time, Zach and I lived there alone, and we had a chain link fence in the back yard. We transferred the bear into the more open crusher-screen-built so-called circus trap so he'd have fresh air and we could easily clean it every day with a garden hose. We parked it in the shade in the back yard, padlocked the door shut so some prankster couldn't release it and waited for Dr. Nelson to let us know he was back in the United States. Zach, who was about 12 at the time, took charge of caring for the bear, which he affectionately dubbed Killer.

One day Zach was invited to play ball with some friends after school. "Sure," Zach said, "but I've got to go home and feed Killer first." The word got out, and before long, everyone wanted to see Killer. For the next two weeks Zach was constantly giving tours to his schoolmates and telling the story of the "dangerous bear." Adults arrived as well. One man who lived north of us in Saxon Harbor liked John's plan so well he donated money to feed Killer and occasionally dropped off sweet corn for him. Zach would push the ears through the grates and Killer would pull them in and eat them.

In hindsight, I think Killer may have gotten a bad rap. All animals have a zone of tolerance, a boundary marking their "personal space." Think about people: we comfortably talk to each other from about two or three feet away, but try talking to someone at about half that distance. They will back away in discomfort. Add a little intimidation or fear to the situation and you have conflict on your hands. With wild animals that zone is much larger. Even relatively timid animals like woodchucks or beaver, if cornered, will charge. I'm wondering if, when Ed Tafelski rounded the corner of the pig pen with the pail of pig feed and met Killer face to face,

Killer attacked out of surprise and fear at being cornered in a tight space. But of course, I could be wrong.

Towards the end of the two weeks, we got the go-ahead to transport Killer to his new home. Dr. Nelson was back, and the Carl Foundation wanted us to drive Killer to Chicago, even offering to pay for our time and mileage. The plan was that John and I would drive the bear to Milwaukee and transfer it to Dr. Nelson's van. They would then drive Killer on to the lab and ship the circus trap back to us by freight truck a week later. Off we went, with food and water stored in the truck to keep Killer comfortable on the trip. It was uneventful until we got on Interstate 94 south of Portage.

People just weren't used to seeing a caged bear going down the highway. In a state truck we couldn't exceed the speed limit like everyone else, so cars on the four-lane would pull up to pass, see the bear, and then slow down for a longer look. They'd parallel us for long stretches, backing up traffic. We finally got to Milwaukee and met Dr. Nelson, who informed us that Killer would be in an enclosure with a female bear from Colorado. I thought, what a happy ending, if Killer is getting life in jail, whether guilty or not, at least he gets to share his cell with a girlfriend.

Months later John got a letter from Dr. Nelson telling us Killer had been renamed. He enclosed a newspaper article describing how the Carl Foundation had started a naming contest for him. Children visiting the Foundation had submitted ideas for a name-the-bear contest, and Killer was now Alfonzo.

But the story didn't end there. A few months later we heard from Ed Tafelski. Apparently, Ed's dog Bandit was getting some overdue attention for his brave defense of his master. He'd been nominated for Kennel Ration Dog Food's "Hero of the Year" award.

The Loon Day Bear

The town of Mercer, population 800, has a big summer event called Loon Day. It's the first Wednesday in August and has been going on for close to 30 years. Back in the 1980s, John Olson had been approached by the Chamber of Commerce president, Daryl Huber, about a town mascot. Our neighboring town, Park Falls, was dubbed the "Ruffed Grouse Capital of the World." Other nearby towns included the "Black Bear Capital" (Glidden) and the "Musky Capital" (Boulder Junction). Daryl wondered what John thought about Mercer becoming the "Loon Capital of the World." John did some checking and discovered a town in Minnesota had already claimed that title. But Mercer could claim the "Loon Capital of Wisconsin."

We were pleased, especially because loons were different from the other town's mascots—all of which were game species, ones you hunted or fished for, and eventually ate. Loons represented wildness, nature, beauty, clean water and air. They had evocative calls, from the wolf-like "wail" to their trilling "tremolo." Loons had charisma and mystique, and we hoped we could pass some of that on to Mercer.

Mercer is located on a historic Native American portage between lakes, and it's common to be walking around town and hear loons calling.

Soon after Mercer declared itself Wisconsin's Loon Capital, Loon Day was established to draw in tourists. The festival features a loon calling contest, where kids of all ages compete by performing their best loon calls. There are also the usual cadre of food vendors and craft tents you would expect at any small-town event. In its early years, Loon Day had a great parade that followed Highway 51, the main route through town, and ended at the DNR Ranger Station parking lot.

In those days the DNR's various bureaus worked together a lot. If there was a forest fire, for instance, we all helped out, and if we (in wildlife management) had a project such as erecting an osprey platform, fish management and fire control staff would lend a hand. Friday nights after a week of working together often ended with our crew sitting in a local bar sharing a beer together before heading home.

Jim Cox was the fisheries technician and a good friend of mine. In the 1980s Jim was single, like me, and renting a house near Mercer. His parents had come from Ohio to visit him for a week, bringing along Jim's sister Kathy, who was 18 and had graduated from high school that fall. Jim had told me Kathy was considering majoring in wildlife ecology in college but hadn't decided for sure.

She wanted to learn more about the role of a wildlife technician and biologist. Jim had told her about my bear trapping and she wanted to see a bear. If, by any chance, I caught a bear while Kathy was there, could she ride along? It just wasn't the time of year bears were much of a problem—there was a lot of natural food available and they were behaving themselves. As luck would have it, I only had a single trap set that week. On Wednesday morning I got a call from the landowner near Ashland where the culvert trap was set. I called Jim and asked if Kathy wanted to come along. I met her at the station and she was excited. We hooked up the trailer and headed out. When we arrived at the site, we took a flashlight and looked into the trap. Gosh, was it a beauty, a huge boar.

Bruce Kohn's research in the 70s had taken measurements on 312 bears, and his average weights were 125 for females and 165 for males; but males can exceed 700 pounds. This bear probably wasn't a new record, but he was the biggest bear I'd ever caught. He was in my biggest culvert trap and filled up the whole trap and couldn't turn around inside. I estimated he weighed 500-600 pounds. His head was enormous, and he was so heavy that, if we processed him on site, I feared the two of us wouldn't be able to pull him back into the trap for the ride to the release site. So we decided to haul him back to Mercer.

We winched the trap onto the trailer and headed home. It was Wednesday—in fact it was the first Wednesday in August, Loon Day. When we arrived at the Mercer station things were busy. Loon Day attracts over 10,000 visitors to town. John Olson was downtown judging the loon calling contest, as was the warden, Jack Kunath. Rosy Richter, our secretary, was in the station answering questions and handing out brochures to tourists. The fire control guys were on the Smokey Bear float in the parade. But luckily Jim Cox was in the shop working on an outboard motor along with two others from the fish crew, Jeff Richter and Tom Cawst.

We pulled up in a corner of the parking lot to avoid attracting attention. The fish guys came over, and I decided to tranquilize the bear right there and have Kathy put in the numbered ear tags. The fish crew would help get the tranquilized bear out of the trap and back in again, so we could release him on public land out of town.

I estimated the bear's weight, calculated the dosage, loaded the syringe and delivered the drug through the hole in the side of the culvert trap. I had estimated what I thought was a fairly large dosage of ketamine, but it's easy to misjudge the weight of a large bear. They can hide a hundred pounds very easily. The pole syringe delivery went well. I hit the spot—how could I miss on a bear this big? I gave him the dose for a 500-pounder and we quietly waited the seven minutes for him to go down. I checked for tongue licking and head bobbing, and after seeing these symptoms we waited a little longer. I didn't want to take any chances. I had the rest of the bottle of drug next to me and another syringe. Kathy was ready with the ear tags and applicators in her hands, but the bear's head was still up. He was sitting kind of hunched up in the back of the trap next to the door, and I was a little puzzled as to why he wasn't down yet. I'd never given a bear this large a dose before, but I decided I'd better give him a booster dose, just in case.

Because he was hunched against the door, the easiest delivery was to just lift the door about 10 inches and give him the shot by hand directly in his hip. When I did this, he didn't jerk or move, and I noticed when I withdrew the needle some of the drug squirted back out. It was as if his fat was so thick and his muscle so tight in his hunched position that his muscle wouldn't absorb the drug. "That's interesting," I thought, "never seen that before." I waited a few minutes to let the second dose take effect and then gave him a gentle poke. He did not move so I thought he was surely down for the count and lifted the door to pull him out. That's when things started happening fast. The bear rolled backwards, right out of the trap, got his legs under him and on four shaky legs started wobbling down the side road toward town.

Kathy screamed, dropped the ear tags, jumped in my truck and locked the doors. Rosy, who'd been watching from the ranger station window, ran outside and shuffled some tourists into the station. Jeff and Jim ran to the shop, grabbed two shovels, then ran back out chasing the bear. I tried to get into my truck to get another bottle of drug from the kit, but Kathy refused to open the door. I finally gave up, dashed the 50 yards to the station, through the back door, down the stairs, grabbed another bottle of ketamine out of the refrigerator, and then, panting, ran back to the scene.

When I caught up to the bear, the fish guys were walking alongside him, guiding his movements with their outstretched shovels. He was wobbling down the station driveway that led to the highway, weaving like a drunken sailor. That's when I heard the drum music coming from the Mercer High School Band. The band was leading the Loon Day Parade. In less than five minutes the flow of people would be marching up this same street.

We had to act fast. While I loaded another syringe with a third dose, the shovel guys turned the bear onto a side road that led into the woods. The bear entered the trees, started up a steep grade, then lay down. I walked up and gave him another dose. That did the trick. He didn't get up again, and we waited as the parade passed around the corner and the band faded away.

Jeff and Jim stayed with the bear. I walked back and talked Kathy into unlocking my truck, and then we parked the trailer as close to the bear as we could. We put ropes around the bear's back legs and pulled it 30 yards to the trap. It took all of us to move that bear. It never did fully go down from the drug and would reach out with his front feet and grab trees to resist us as we tugged it along. Finally, we got him tagged and back inside the trap. We all sat down, exhausted. Rosy's tourists emerged from the station to have a look.

Kathy and I drove the bear to a nearby business, Saarnio's log yard, where we weighed it on a scale used to weigh logging trucks. After the bear was released, we drove back and weighed the trap and trailer again, subtracted the numbers, and found our bear weighed over 650 pounds.

The last thing that went wrong that day was that while we were transporting the bear to the release site, the small bait door in the front of the trap broke and came open. The bear stuck his huge head through the door and got stuck. He was still a little drowsy from the drug and couldn't pull his head out. We had to tap him on the snout and turn his head on its side to get him free.

Kathy went on to college that fall as planned but changed her major. Wildlife biology, it turned out, was no longer her first choice.

Loon Day is still a big event in Mercer each year. If you ever drive through on Highway 51, be sure to stop and look at the 16-foot tall fiberglass loon located next to the Chamber of Commerce. Its name is Claire...as in, Claire de Loon.

The Stanchion

The memo came from district headquarters in Spooner, the office covering Wisconsin's 12 northwestern counties. It stated that starting immediately, all black bear chemical immobilization must cease by the first of August each year. I wondered why. We generally caught a lot of bears in August; that's when the majority of corn damage occurred. I could trap and relocate bears without drugs, but how was I going to ear tag them?

Bruce Kohn had always stressed how important our tagged bears were to his population modeling. Some years our northern wildlife staff ear tagged an impressive number. One year our crew at Mercer caught 88 and Dick Smith, the wildlife tech at Brule, caught over 100. Because these ear-tagged bears were spread out over the range so evenly, Bruce said, he could calculate the population by knowing how many were tagged and how many tagged bears were recorded shot during the hunting season. Here is an example of how the math works. If you had a 100-square-mile study site and caught and tagged 100 bears within that area, then during hunting season, hunters harvested 100 bears in the area, of which

10 were ear-tagged, that would mean you'd harvested 10 percent of the bear population, so the total population had been 1,000 bears. This technique is called mark and recapture. Fish management uses a form of it to determine fish populations in lakes and streams.

This was not Bruce's entire population census technique, but it was an additional measure he could track to make sure our model numbers jibed. It's not easy to count wild animals. They don't line up along the road on a certain day for roll call. Along with this data, ear-tagged bears helped convince farmers that my trapping efforts were successful and I was alleviating their damage problem. I sometimes caught several bears of similar size in an oat or corn-field. Farmers would often say, "That's the same bear you caught last week, he just came back." I would remind them, "Don't you remember you helped me ear tag last week's bear? This one doesn't have a tag." I think I saved the lives of a lot of bears this way. Most farmers requested, and were issued, a damage kill permit. If they thought I was eliminating their problem by live-trapping and translocating the bears, they didn't shoot them.

Well, I lost the argument about stopping use of tranquilizer drugs on August 1. It seems this order came straight from the Federal Food and Drug Administration (FDA). Livestock producers are not allowed to administer drugs, vaccines, or hormones to any meat producing animal 30 days prior to slaughter. Similarly, they didn't want people eating bear meat with drug residues in it. Bear hunting season opened in September. Their reasoning made sense in that context.

A week or so later, I was driving down the road with a bear in a culvert trap when I remembered the Loon Day bear and how he stuck his head out the broken bait door trying to escape. The door was 12 inches wide and 18 inches tall, too small for him to escape through but he tried anyway. That reminded me of something from my childhood back on the Iowa farm. When we milked cows we secured them in a stanchion, a metal frame that held the cow stationary by the neck so we could milk them without being kicked. Could a bear be stanchioned like a cow?

When I got back to the station, I approached Dave Sleight. He was a fire control equipment operator and was the best welder and metal fabricator I knew. He was always inventing things to use on his bulldozer and fire plow. He invented a fireproof case for mounting emergency fire shelters on the fire plows and they are now standard equipment on all firefighting units in Wisconsin.

I explained to Dave how important ear tags were to our population model and my idea of mounting a cow stanchion on the bait door end of a bear trap. I knew if it was ever going to happen I would need Dave. He looked at me and said, "Are you crazy? That's the stupidest idea I've ever heard. I want nothing to do with it."

I started bugging my boss, John Olson, but he was on Dave's side. But I am nothing if not persistent, and John must have gotten worn down because one day he showed up and handed me a cow stanchion. He'd been home to his family's farm in Winter, Wisconsin, that weekend and unbolted it from his dad's old cow barn.

A few weeks later I was helping with some forestry work in northern Iron County at the Upson Ranger Station. Rudy Kangas, who was stationed there as the fire control equipment operator, had an 80-acre farm south of the town of Iron Belt, not far from Upson. I'd occasionally drive past Rudy's farm and had noticed some old, abandoned, broken-down farm machinery he'd left in the corner of one of the pastures. I told Rudy about my stanchion idea and that I needed a few more parts. Would he mind if I checked out his machinery dump? He just grinned, puffed away on his pipe, and said to help myself. It wasn't long before I was combing through that old machinery and found the part I needed. It was on an old hay rake—the lever and gear that raise and set the height of the rake—just what I was looking for to hold the stanchion locked shut on the bear's neck. It even had what would serve as adjustable notches for different-sized bears. A half hour later I was heading down the road with that lever and gear in the back of my truck.

Somehow, I had to convince Dave Sleight to help me, so I started bugging him again. I was lucky because it was midsummer and the woods had greened up, fire danger was down, and Dave had more free time. He finally gave in. The one thing he insisted on, though, was that when the contraption was built and installed on the trap he didn't want to be around when I tested it.

The day finally came and we had the stanchion mounted on the circus cage trap. We had picked this trap because the mesh sides let in a lot of light so we could easily observe the bear inside. I had a bear in a culvert trap that we could use to test the stanchion, so we placed the circus trap door to door against the culvert trap that held the bear and lifted both doors at once. The bear ran into the light-filled circus trap thinking it was a way out. We lowered the door, enclosing the bear inside the circus trap and then I opened the stanchion door. The bear ran to the stanchion and stuck his head through, thinking he was escaping. I quickly locked the stanchion down on his neck and had him secured. All of the Mercer crew was there for the demonstration, except Dave of course. The bear pulled back and turned his head sideways and easily slipped its head out of the stanchion. It didn't work, my design had failed. We went back to the drawing board.

Dave had some ideas. Instead of mounting the stanchion vertically like a normal cow stanchion we mounted it horizontally. Next, we welded a V-bar on the bottom where the bear's chin would sit so the bear couldn't escape by turning its head sideways. We also shortened the stanchion and welded a wire mesh panel on the lever so we could refine the adjustment of the hole around the bear's neck. Prototype two was ready for testing. We just needed another bear.

A few days later I caught one, and when I pulled into Mercer station everyone came over for the demonstration—the warden, fire control ranger, fish crew, foresters, everyone except Dave. It was a nice-sized bear, maybe 200 pounds or more. We repeated the trap-to-trap maneuver, and the bear walked into the stanchion trap. I had placed a small tarp over the stanchion so he wouldn't see me standing against the trap. The bear saw the open hole of the stanchion and stuck his front leg through, drew it back, and stuck his head in. I made my move and clamped down the stanchion on the bear's neck and then stepped aside. The bear pulled backward against the stanchion. He turned his head sideways and pulled back again. This time he was held. I picked up my ear-tag pliers, carefully reached down from above, lifted up an ear, and clipped on a tag. The bear tried to bite me but couldn't lift his head. I repeated the same procedure on the other ear, released the locking lever, and the bear sat back on his haunches, looking at us with his two new shiny white tags. I only wished Dave could have seen it.

John and I stanchioned dozens of bears over the next few years. One fall, we must have stanchioned over 30. We got so good at it we could do it alone, and I once stanchioned three in a row in about 20 minutes. Margaret Levra, from the Ironwood Daily Globe, came down and wrote a story about it. Somehow a Green Bay TV station got wind and sent a crew to film it. Colleen and I stanchioned and tagged a big 300-pounder for them. I gave a demo at our annual district wildlife meeting, but somehow none of the other wildlife technicians took any interest in trying it. I continued to use it as long as I worked on bears, but as of this writing that old stanchion trap is sitting back behind the equipment shed at the Mercer DNR station. I always felt a little nostalgic when I walked by it—after the resurrection of its parts from Rudy's field and Carl Olson's old barn as perhaps the world's only bear stanchion, it had returned to retirement and was rusting away with weeds and brush growing up through its iron mesh.

The Feds Take Over

The bear trapping years ended at a district meeting in Spooner. Bruce Moss, the District Wildlife Specialist, told us the Madison Wildlife Bureau had been talking to the federal Animal Plant Health Inspection Service (APHIS). APHIS had been working in Wisconsin on Horicon Marsh, doing goose damage abatement work because geese are migratory and federally regulated. Bruce explained that now that APHIS had offices in Wisconsin and some personnel, they had given us (DNR) a bid on the bear control work. The Madison staff checked the cost, and it would be higher than what we were doing it for, but they offered to cost-share with federal money so it would reduce our overall budget. Besides costing our wildlife bureau less, it would free up our technicians to do more valuable work such as wildlife habitat projects. Bruce felt it was a good idea but wanted full support from the field staff. I knew this was coming as I had heard about it and was dead set against it. An upper-level management meeting had been held and a district meeting was set up so all the wildlife personnel could discuss the matter.

I decided to speak up. I felt that if we were the state's Bureau of Wildlife Management, then we should manage the bears. They were not federally managed like migratory birds—they were our responsibility. We surveyed them, we set harvest quotas, we registered them during the harvest, and we did the population modeling. Why would we want another agency doing control and abatement work? Besides, I stated, after years of trial and error we were good at our jobs. We had the experience, the equipment, and the knowledge. They didn't.

Bruce Moss interrupted me: "Now, I don't want to hear any of that 'ain't broke don't fix it' mentality." Then I stated that we had a good rapport with our farmers, relationships that took years to develop. I said that some years we had very little bear damage work—for example, one year I only caught seven bears, another year it was 88, but we averaged about 20 or 30. What were the APHIS employees going to do on those low complaint years? Sit around and twiddle their thumbs like the guy in the Maytag washing machine commercial? I went on to argue that when I had a low bear complaint year, I had 20 other things to do in my job description. This new system might cost our bureau less but so what—it would cost the taxpayers more. As Bruce had just said, it was either us or the Feds, but either way it was taxpayer money. I suggested he was missing the big picture and ignoring the interests of the citizens.

I told them what Harold Schmude, the warden at Pence, had told me when he turned over his bear traps to me years ago: "I didn't want to give up bear trapping. I was forced to do it. Bear damage control was the one thing we wardens could do for the public that wasn't law enforcement. The public thanked me a lot for it. You think they thank me for writing them a ticket for violations? It was one of the best public relations services a warden could do. I made a lot of friends who later became informants when violations occurred."

I summed up, saying that the public and the politicians look at bears as dangerous animals. Someone will always have to be available to respond to bear problems. Bear control work secures the wildlife technician positions. Lastly, I liked the work. I enjoyed what I did. The vote was mixed but others supported my points.

The decision was made a few weeks later. The Feds would take over bear control duties. We'd lost the bear damage control program just like we'd lost the beaver control program years earlier. I believe that this was the beginning of a change in personnel that took many wildlife staff out of the field, weakened the security of field techs positions, and reduced the public visibility of our work.

About a month later John Olson called me into his office to tell me he had just gotten a call from Spooner requesting my help. The Feds were having a training session at a campground near Rhinelander. They had hired new personnel to be their bear trappers and the DNR wanted me to teach them how to trap, foot snare, tranquilize and ear tag bears. I went, actually taking a bear in a cage trap with me to do the demonstration. I'm sure it was the first bear many of the federal trainees had ever seen. Later the federal trapper covering our area came to the Mercer station, and I gave him all my traps and equipment.

Today, APHIS does all the beaver, black bear, Canada goose, and wolf abatement trapping in Wisconsin. I'll be the first to say they do a great job, but unfortunately most of the wildlife technician positions in Northern Wisconsin have been cut 50 percent or more. The DNR stations at Park Falls and Brule that I worked out of in the 1970s, 80s, and 90s each had three wildlife technician positions. Today, they have none. After I retired, my position in Mercer was not filled. Political changes and subsequent budget reductions played a big role, but I think if they'd kept animal damage control work, especially the bear nuisance and crop damage abatement work, we'd have had a strong argument to keep the wildlife technicians, and to my mind, we would have kept stronger community ties as well as ongoing wildlife education and research. Ironically, at the time of this publication, I handle dozens of beaver complaints each year as a private contractor helping townships and county government because APHIS can't keep up with the workload and the DNR has no staff to assist.

Osprey Penthouse Apartments

"The last word in ignorance is the man who says of an animal or plant, what good is it?"

—Aldo Leopold

Much of my work in the late 1970s and early 1980s was focused on forest habitat projects including aspen forest regeneration, maintaining and creating grassy wildlife openings, animal damage control, and wildlife population surveys. This was a time when wildlife management was beginning to shift from game management, which focused solely on species humans hunted or "harvested," to a broader ecological approach that recognized the importance of biological diversity and "keeping all the parts" of the ecosystem, as ecologist Aldo Leopold famously argued in his book *Round River*. He wrote, "If the land mechanism as a whole is good then every part is good, whether we understand it or not...To keep every cog and wheel is the first precaution of intelligent tinkering." At our northern Wisconsin ranger stations, our job titles had long ago changed from game management to wildlife management, but most of our work was still targeted on game species. After all, most of our funding came from the sales of hunting and trapping licenses.

The DNR's first fledgling office on endangered and non-game species was started in 1978, but it was underfunded and understaffed. Things improved somewhat in 1982 when the Bureau of Endangered Resources was established. Much of the fieldwork on non-game, and threatened and endangered wildlife projects was done by enthusiastic wildlife management personnel after their game management projects were done. When we attended conferences or statewide professional meetings, new theories and approaches would be discussed, and we would come home filled with ideas. John Olson and I understood that the field of wildlife conservation was evolving toward ecosystem management and, increasingly, we felt non-game wildlife had a place in it. Perhaps, too, we saw opportunities to expand our horizons and have some excitement learning to work with new species.

John and I had attended a presentation by Ron Eckstein, the assistant wildlife manager in Rhinelander at the time. Ron, with his technician, Fred Johnson, had been constructing artificial nesting platforms for the osprey (fish hawks) in Vilas and Oneida Counties. Osprey were then a state-listed endangered species, and like eagles, had been negatively affected by DDT and other organo-chlorine chemicals, as well as the loss of nest sites and death from illegal shooting. The DDT issue was partially solved by banning its use throughout the United States in 1974, but the number of birds was still reduced below what was deemed a recoverable level. At that time Wisconsin had only 82 breeding pairs of osprey.

The department conducted two aerial surveys each year, the first to count nesting pairs of osprey and eagles, and the second to measure productivity by counting chicks. Results showed both species' populations were slowly increasing. Ron was having great success boosting osprey populations by installing nesting platforms, as they could be tailored to fit their unique nest requirements. Unlike eagles, who prefer shade and usually nest further down in the tree's canopy, osprey require a nest site at the top of the tallest tree in an area. They prefer to nest near water because they only eat live fish, whereas eagles eat live fish, dead fish, carrion, small mammals, waterfowl, road-killed deer—you name it. Osprey are also fussy about the neighborhood. They prefer that all the nearby trees are shorter, avoiding locations where eagles or other predators can find a higher perch and look down at their nest and chicks. Lastly, osprey are not very good at expanding their home turf. Wisconsin osprey migrate to South America in the winter, and young birds follow this same path. When these young birds return to the nesting grounds as two-year-old adults, they generally settle within 10 miles of their original birth site.

For osprey, finding a location with all these nest characteristics can be quite rare, and even when a perfect site does exist, they face challenges. Building a large nest on the top of a live tree is almost impossible because the terminal leader (the top branch reaching skyward) keeps growing upward and is thin and weak. So most natural osprey nests are built in dead trees. Dead trees, of course, have a limited time they can support a heavy stick nest or withstand wind and the force of gravity because, after all, they are dead. Also, the tops of dead trees are difficult places to securely construct nests, as branches often break off and windstorms can blow the nests apart.

Ron's platforms looked like a solution to the osprey nest site problem. They could be placed in live, super-tall red and white pine trees to give them the needed height and stability, we just had to trim off the terminal branches at the top. We could select trees near water bodies with good fish populations and place the platforms within 10 miles of existing osprey territories based on our aerial survey data. We jokingly referred to it as moving the osprey from slum housing to luxury penthouse apartments.

After hearing Ron's talk, John and I felt we had the ideal situation in Iron County, excellent lake country with a scattering of natural osprey nests already established. We also had the Turtle Flambeau Flowage with its 14,000 acres of water, 95 percent wild undeveloped shoreline, and a history of osprey nesting. Flooded in 1926, the flowage once had thousands of dead standing trees that over time had succumbed to wind and decay. Osprey had used those dead trees to nest in the area before time toppled them. We believed we could increase the birds' success by beefing up their summer housing options.

Through correspondence with Ron we got copies of his design—a three-foot diameter hexagonal platform frame constructed from two-inch thick lumber. Chicken wire formed the bottom, and the platform was attached to the tree with three iron straps on the sides and a spike down the middle into the top of the tree. Bob

Baldauf, from the DNR fire control crew, built our platforms in the ranger station shop during the winter, while John and I studied historical locations and maps and reconnoitered trees from the ground. We tested our first platform on a tall pole once used to support an old fire control tower telephone line. We attached the platform to the 25-foot pole, cut a hole in the ice, and with a crew of five people, attempted to secure it by pounding it into the soft, mucky bottom of the flowage. The pole was wobbly and almost fell on us as we struggled to raise it. After doing it once we decided to go back to the drawing board and try Ron's technique involving live trees. We borrowed "pole climbers" from the fire control crew. These are metal spikes you attach to your boots with leather straps. We then put a climbing harness around our waist with a flip belt about four feet long. This belt wrapped around the trunk and secured us as we made our ascent. By jamming the spikes into the tree and flipping the belt up with each step we could inch our way up the tree. Two problems arose. First, the climbing spikes were too short. They were designed for telephone poles, not 30-inch-diameter pines with thick bark. They barely scratched the surface, let alone penetrated the bark to enter the cambium layer where we could get a good secure grip. Next, the flip belt was great until you got to a limb. Sometimes the first limb was 30 feet or more off the ground, making it very dangerous to detach the belt and reattach it above the limb before proceeding up. The eagle banders, Chuck Sindelar and Ron Eckstein, took pity on us and gave us their extra tree climbers, which had long three-inch spikes that penetrated effectively. For the limb problem we added a second belt and simply attached the extra one above the obstructing limb, then detached the lower one and proceeded upward. After we acquired a lot of experience, we just climbed with our fingers holding the bark and didn't strap in until we got to the top.

I remember the first tree I climbed solo. It was a nearly dead white pine about 30 feet tall with no limbs, located in the middle of a beaver pond on the Turtle Flambeau Flowage. From the ice I "flip-belted" to the top with no problems, using my sharp climbing spikes to stabilize myself as I went up. Next, with my rope I pulled up the heavy chainsaw, somehow started it and cut the tree off about a foot from the top with the chainsaw bar cutting just inches from my face. I then lowered the saw and pulled up the tool bag and tied it to my waist. The platform came up next and I don't know how I found the strength to lift it onto the treetop. I hammered in the center spike and then screwed the lag bolts to strap the platform down. Finally, I pulled up a clump of sticks and arranged them in the platform to resemble a nest. By the time I'd climbed down I was so shaky, cramped, and exhausted I was done for the day. My entire body hurt. I'd been holding myself for over an hour with my legs on those two pole spikes.

As we gained experience, John and I perfected our technique. One of our first improvements came when our seasonal wildlife employee, Charlie Zinsmaster, taught us how to rappel down from the tree using a figure-eight descender. This was the same device used in Vietnam to quickly get soldiers out of helicopters and down a rope under enemy fire. It saved us from having to free-climb down the tree when we were done, as a safety rope was difficult to rig without the descender. A second improvement came from a friend, Gary Mears from Mercer, who had been helping us for a day. He was a sailor and had a very practical eye for the tasks at hand. He said, "Why don't you attach a pulley to the top of the tree and have your ground crew do all the work of pulling up your chainsaw, tools, platform, and nest material so you can save your strength?" John, Charlie, and I looked at each other... YES, and a pulley was quickly found and put into service.

After about three winters of osprey platform placement in Iron County, our data started coming in. When combined with Vilas and Oneida County numbers, osprey in our three-county area showed remarkable success. We had expanded the TFF population from five to 22 nesting pairs, supporting our theory that nesting habitat was a limiting factor for Iron County osprey. Osprey platforms also averaged higher chick fledging success than natural nests. We surmised that was because they were more secure structures and withstood windstorms better than natural nests.

This was proven to me firsthand on a couple of occasions. One time in the 80s, John Olson got a call from a resort owner reporting that a spring windstorm had blown over a nest tree just after the birds had returned to nest. It was a natural nest over water out on the TFF, and Ken Neuman, the resort's owner, asked if there was anything we could do. The nest was visible from his resort's big window in front of the bar and the birds' activities were quite an attraction for his customers.

We located a giant white pine on the opposite shore and put up a platform. After we boated back to the resort landing, Ken offered us hamburgers and cokes. We had just settled down to begin our meal when Ken pointed out the window. Following his gesture, we looked out and saw two osprey land on the platform with sticks in their talons to start building a new nest. John looked at his watch. It had only been 20 minutes since he had rappelled out of the tree.

A second example occurred years later when an occupied osprey platform on Little Turtle Flowage was struck by lightning. The top of the tree was blown off and at its base we found one dead chick and one emaciated, but live, chick. Bruce Bacon rushed the chick to a rehabilitation center, and we cut the tree off about 10 feet below the original platform and installed a new one. Five days later the chick was returned from rehab, so I climbed the tree again and placed it in the new platform nest. Within 15 minutes the adults were delivering fish to their prodigal chick.

After other wildlife managers heard about our success with platforms, many wanted to erect platforms in their counties. We were then part of the 12-county Northwest District and we thought that the lake country around the towns of Hayward, Spooner, and Brule had great potential. The national forest biologists were interested as well so John gave presentations about our project at their district meetings, and Charlie and I gave climbing and equipment demos. By then, we were only doing the work in the winter as it gave us better access to wet marshy areas by crossing the ice with snowshoes and snowmobiles. We were commonly climbing up 90-plus-foot trees and could erect up to four platforms per day if access was good. At that time, we were putting in long days, sometimes rappelling out of our last tree in the moonlight.

Eventually we took our osprey platform dog-and-pony show on the road. The local wildlife managers and technicians would pick out potential platform trees, get permission for access, and do all the groundwork. John, Charlie, and I would hit the road for a week at a time putting up platforms. We always offered to train local staff but for the most part they showed very little interest. I guess they just couldn't grasp how fun it was to climb a frozen tree in the middle of the winter, run a chainsaw inches from your face, and lift heavy objects while hanging 90 feet in the air from a half-inch rope. We loved it (mostly), and often told them not just anyone was qualified to do this work, you had to take a test. The punchline: it's the common IQ test and you have to score low to get the job. No one is more qualified than us. We had some great times. We brought sleeping bags and camped out on cabin floors, in hunting camps, and ranger stations. We installed over 200 platforms in Wisconsin, from the Peatenwell Flowage in central Wisconsin to the shores of Lake Superior. We went to the Chippewa National Forest in Minnesota and to Lake Macbride in Iowa, where we also sent osprey chicks to establish an Iowa population.

Since our efforts in the 1980s, platforms have been erected all over the state. This work has been done by the DNR as well as a multitude of cooperators including electric power companies, Native American tribal nations, private environmental education centers, the National Forest Service, conservation clubs, and private landowners. Much of our work and the work of others was funded by the Bureau of Endangered Resources.

Today, osprey in Wisconsin have been removed from both the state endangered and threatened species lists. The population of bald eagles has also recovered. In 2016 the state reported 1,500 bald eagle nests and 558 osprey nests. Our platforms are still popular housing choices for the state's osprey, with 75 percent of the nests on artificial platforms. In fact, our high-quality housing even appeals to eagles. When you see an artificial platform today, you should look twice to see which of these impressive raptors is peering down.

A Close Call

One winter in the late 80s, John, Charlie, and I were on the road for a week of osprey platform installation. It was early morning and we were headed to a remote lake in the Pigeon Lake Wilderness Area of the Chequamegon Nicolet National Forest. We always scheduled fieldwork months in advance to line up sup-

plies and make sure we had a ground crew available. However, this meant we didn't always have perfect weather conditions when we arrived. John and I would rotate as climbers because climbing took a lot of energy. This morning it was my turn to start first. The weather was a bit colder and windier than I preferred, and the tree bark and cambium layer under the bark was frozen. I knew this before we left our motel and, when dressing that morning, tried to calculate the right clothing for the day. What I didn't know until we met up with the Forest Service personnel was that I had a two-mile snowmobile ride to get to the site. If you haven't done fieldwork in the winter in the northwoods, you might not be familiar with the challenges and equipment needed to travel to remote locations and work in deep snow and frigid temperatures. A snowmobile is essential for any lengthy trip, and we would often haul a sled behind the machine. This "musher sled" would allow us to tow heavy gear like platforms and saws, and another person could ride behind. I didn't like snowmobiles and preferred to ride on the musher sled with the equipment. The sled was the dog-sled style where you stood upright on the back with your feet on the runners. The tongue of the sled attached to the snowmobile with a hitch that flexed with the terrain.

When tree climbing in winter, you want your clothing to be flexible so you can swing your legs and arms up and over branches and jam in your spikes at odd angles. You also want them warm and absorbent so if you perspire you don't get too cold. I hated overdressing for climbing because I usually used so much energy during the climb that I was a sweaty mess when I got to the top. If I had to wait too long in a strong wind for the equipment to be pulled up, I sometimes got a chill.

After the two-mile ride to the lake I was starting to shiver. But I felt okay and didn't mention it to the others. The Forest Service biologist, Joyce Zifko, pointed out the tree she had picked. Boy, was it a beauty—a huge white pine standing on a hummock in the middle of 100 acres of tag alder brush. Due to its location in the swamp, it had to have been one of those rare trees that escaped the saws and axes of the pine-logging era of the late 1800s. We bushwhacked a snowshoe trail up to its base and carefully looked it over. Then we turned to Joyce.

"Well, you told me ospreys want to be up high," she said.

The tree had to be four feet in diameter, maybe more. It was about 80 feet tall and bushed out at the top, probably from tip weevil damage, a disease white pines can get if they grow in full sunlight. I'd have to saw it off about 10 feet down from the top so the platform would be above the limbs.

I realized right away that the trunk was too big in circumference to get my flip belt around, so I'd have to free-climb, gripping the bark with my fingers. It looked to be about 30 feet to the first limb. Then it had limbless segments of 10 feet or more where I'd have to free-climb again before it narrowed enough that I could wrap my flip belt around it. This was a challenging quirk, as usually once you are in the canopy of a white pine, you have abundant limbs for handholds all the way to the top. The tree was so wide it didn't feel round. This climb would be like scaling a 70-foot wall of frozen bark without ropes, or belts, or limbs to catch a fall.

I strapped on my climbing spikes and harness. attached my haul rope, and started up. I was still shivering, slightly. But this didn't bother me because I knew I'd warm up quickly once I got into the climb. Right away I could see gloves wouldn't work because I needed to feel the bark grooves to get a firm grip. When I jammed my spikes into the bark, it was so thick and frozen it felt like stone. My spikes barely scratched the surface, coming nowhere close to stabbing into the cambium layer. Oh well, it wasn't my first tree. I'd climbed many like it, and if I could get above that first limb 30-foot up, I knew I'd be fine. So up I went.

When climbing free-style, all you need are three solid points of contact. That means two legs and one hand, or two hands and one leg. As long as one of those three points doesn't slip, you're fine. Another concern is fear—you can't panic. That makes everything worse.

That day, I had no fear, or not much. The first 10 feet were easy. The next 15, no problem, the same. About five feet below that first limb, I froze. I don't mean froze in the literal sense of cold—my arms and legs just wouldn't move. My brain told them to work, but they didn't. I could talk, but my body wouldn't respond to my commands. John had been watching closely from below. He knew

I was in trouble and started talking to reassure me. I'd had a big breakfast that morning, was in great physical shape, and had been climbing every day for the past several days. I had no idea what was happening to me. John started clearing people and equipment from the area below me, fearing a fall. I started thinking: "I've got to get to that limb, five more feet." It would be too risky to try climbing down. Climbing down is very dangerous—one slip and you can lose your purchase. Going up, you can jam in a spike, test it with some of your weight, and then give it the rest of your weight, but going down you can't perform that test. John's advice to me in this critical moment was to hang on and rest, so I did. I couldn't do anything else anyway.

After about five minutes I decided to give it another try. This time when I told my brain to pull out a spike, out it came, followed by the next, and so on, as I lethargically inched my way up the tree. I watched my hands and arms working until I felt my head bump the limb. Then, guided by John, I worked my harness belt over the limb and snapped in. I let go and just hung there for 10 minutes or more, resting. My strength and motor control soon returned to normal, so I proceeded to the top of the tree and tied in. In no time, I had the tree topped, the platform attached, and was rappelling down to solid ground.

Ron Eckstein had the answer to what had happened, which my daughter-in-law Melody, a nurse, later confirmed. The medical term is vasoconstriction, and the shivering had set the process into motion. A person's body must maintain a core temperature around the heart, and because I started the climb already chilled, as I used energy with my leg and arm muscles, my heart rate increased and pumped blood and oxygen to my extremities. This lowered the temperature around the core area around my heart, making it react by shutting off the supply of blood to my arms and legs. This caused my arms and legs to refuse to function and made me lethargic. It's the same thing that happens when someone experiences hypothermia and their hands and fingers go numb. The rests in the tree allowed me to warm up and get the blood circulating again.

We had other close calls, but no one ever got hurt. Once, we dropped a chain saw. Every now and then, a platform would detach from a carabiner and come crashing down. Later, when we got climbing helmets, we told people we had so few brain cells in the first place we couldn't afford to lose any more.

Wrong Tree

Working on habitat enhancement projects and monitoring wildlife populations was undoubtedly very important work for our crew. But most of us really lived for those up close and personal moments when we actually handled animals. You can take a child to a frog pond and listen to the peepers singing in spring, see the frogs in the shallows, but catch one of those frogs and place it in the child's hand and see what happens. They are drawn to that frog in an intimate way that makes all future encounters more meaningful—it creates that essential connection to nature so eloquently described by biologist Rachel Carson in her book *A Sense of Wonder*.

In the late 1980s, John Olson got a call from a man named Don Follen, who said he wanted to band ospreys in our area. His call gave us a new dimension for our osprey work that involved direct contact with the birds.

We had built nest platforms, flown population surveys to count the nesting young, and written nest site management plans. I had even been ground crew for colleagues Ron Eckstein, Chuck Sindelar, and Dave Evans, looking on with envy as they climbed massive nest trees to band eagle chicks in their aeries. Now we were going to have the opportunity to have a real wild bird "in the hand." While outwardly maintaining a cool professional demeanor, I was secretly ecstatic at the prospect of this new opportunity.

Don was the guru of osprey banding in Wisconsin. He had a direct connection up the wildlife management "family tree" to Aldo Leopold. Don had banded birds with Fran and Fredrick (Hammy) Hamerstrom, who were Leopold's graduate students and famous for their work with Wisconsin prairie chickens, hawks, and eagles. Like Leopold, the Hamerstroms had developed quite a reputation in the field and mentored many up-and-coming biologists at their research farm in central Wisconsin. Fran dubbed those who worked on hawks and climbed trees "gaboons," and Don was one of Fran's gaboons.

Don wanted to learn more about osprey. At that time, they were listed as endangered in our state, and many aspects of their life cycle were still shrouded in mystery. Don wanted to document chick survival and migration data. He was interested in how many chicks fledge from our nests, how many survive migration, and what routes the birds followed on their southern trips to Peru, Columbia and Brazil. He felt we needed more data on return rates of chicks to natal areas and causes of mortality between their initial flight to South America and their first return to the nesting grounds two years later. Even questions such as dispersal rates—did they nest in the vicinity of their natal nest or show up elsewhere in the state—were foggy at best. A long-term banding project could fill some of our information gaps and help with population recovery. This would require climbing into nests and placing aluminum U.S. Fish and Wildlife Service (USFWS) bands on the legs of nestlings.

We knew where most of our active nests were located, both natural and manmade, but not all would be good candidates for banding. Our first step was to identify which nests were safe to climb and had chicks that year. Don was a USFWS-certified master bander and would provide the numbered bands and train us on how to take measurements and safely handle the birds.

The day before banding John and I carefully went through our gear to make sure everything was shipshape. John was concerned that we conduct ourselves professionally, thinking that if we really impressed Don, he might decide to continue banding with us in the future. John pointed out that I must listen to Don first, then ask questions, and that being near the nest put stress on the birds so we needed to be efficient. John was really laying on the pressure, observing that Don was a certified private bander while we represented the State Department of Natural Resources, and as state employees this was our chance to impress him with our professionalism. We even decided to wear our uniform shirts, the khaki-colored ones with the DNR patch on the sleeve that we wore when giving talks to the public. I had no doubt Don was going to be impressed. Secretly, John and I were as excited as kids on Christmas morning.

The day finally arrived, and John and I got to the station early. We were dressed neatly and had everything in order. We had decided to start on an easy climber, so we picked the nest platform on North Bass Lake for our first tree. It was a few miles east of Mercer and the boat landing access was directly across the lake from the nest. It would be a short boat ride and the tree was within 20 yards of the shore. As we talked the plan through, we decided that because it was our first nest, and because this tree had plenty of limbs at the top to sit on, we'd climb it together—that way there would be two of us if we had difficulties handling the nestlings.

When Don pulled up in an old Dodge pickup with a grumbling muffler, he was wearing faded blue jeans and a brown-stained t-shirt with a bottle of Cutter's insect repellent sticking out of the pocket. He wore a dilapidated ball cap with unkempt curly brown hair sticking out over his sideburns. His skin was tanned and leathery from days spent in the field, and his grin reached from ear to ear. John and I, in our neat uniforms, were quite the contrast. John suggested I ride with Don, and John would drive our truck and tow the boat. We'd do our planning and training session at the landing.

Off we went. Don talked nonstop, drove with his left hand, and was constantly reaching up on the dash for his binoculars if he saw the slightest movement out of the window that he thought might be a raptor. He even drove while looking through his binoculars, and I yelled out once when he veered for the shoulder. The dash of his truck was covered with feathers, skulls, bones and fishing tackle, no doubt things he'd retrieved from osprey nests. He was definitely the real thing, one of Fran's gaboons. I knew immediately I was going to like this guy.

When we got to North Bass Lake, we launched the boat and Don gave us a training session. He told us how to safely handle the chicks, what to do if one jumped from the nest, and gave us a wire hook to reach a chick if it was on the opposite side of the nest and acted like it was about to jump. He explained how to pre-bend the bands so the lock-on part would connect smoothly, how to find and measure the eighth primary feather to determine age, and finished by explaining how to hold and safely replace the chicks in the nest.

As John had suggested, I took all this in and didn't ask many questions. We were looking organized and professional, and John seemed happy. We pointed out the nest tree to Don, a tall white pine barely protruding above a clump of similar pines surrounding it. The nest was about 60 feet up and we could see an adult osprey perched on the edge of the platform, brooding the chicks by shading them with its wings.

We crossed the lake without saying a word. I drove the boat, Don sat in the middle, and John took the bow. When we landed, John carefully got out and tied up the boat. Don got out next, and we passed the gear forward until it was neatly arranged on the bank. John and I slipped into our harnesses and climbing spikes and because John would climb first, above me, we tied our rappelling rope to John's belt. Lastly Don zipped three leg bands and the banding pliers into John's pocket.

John started up the tree, all of us remaining quiet. I waited until he was five feet above me and then started my climb. 10 feet…20 feet…30 until we were halfway. I was full of adrenaline and I knew John felt the same way. When we got to the first limbs the canopy closed in and I could only see straight up. At about 40 feet, I lost sight of John, so I just followed the rope. Then the rope stopped moving and my head bumped something. It was John's rear end, but I knew we still had at least 10 more feet before we hit the top—something wasn't right. John turned and looked down at me. His eyebrows were raised and he looked stunned. Then a sheepish grin appeared and spread from ear to ear. I was baffled. I climbed closer and whispered, "What's the matter?" John leaned down, looked me in the face, and whispered back, "Wrong tree."

I looked over my shoulder at a neighboring pine about 20 feet away and saw two osprey chicks staring at us. I was shocked. So much for impressing the renowned Don Follen with our professionalism. Instead, we looked like a couple of idiots. When we got to the ground John walked over to the correct tree and started up. As I walked past Don, he whispered, "What happened?"

"Wrong tree," I said with embarassment.

"Oh," he whispered. "That's happened to me too."

We climbed the correct tree and banded our first osprey chicks without any further problems. After we loaded the boat and started motoring away, we really lost it, laughing out of control. We couldn't stop laughing about it all day.

John and I banded with Don for the next few years and became great friends. Don could tell a story, and I'm sure this one was re-told by him many times. That day we learned more than to double-check the tree—we never wore our uniform shirts again for osprey work. It was quite a job washing out the pine sap.

Don passed away in 1988. When John Olson later transferred to La Crosse, I became a sub-permittee bander under another master bander named Chris Cold. In 1993, Bruce Bacon became the wildlife manager at Mercer. Bruce is a master bander, and between the three of us we've banded over 500 osprey nestlings in the area. We've received band returns from as far away as Peru. Years later we added numeral-etched color bands and were able to document our own nestling returns to the area by using a spotting scope to read the numbers on live birds.

Hacking Osprey to Iowa

Hacking is a falconry term that describes the method for moving young birds from their natal location to another site to establish the second location as "home." It works very well with some bird species, especially hawks and falcons. The general method is to take a young nestling from its nest before it fledges and move it to a new location where it learns to fly. If done early enough, the young bird imprints on the new location and considers it home. For migratory species, the young will migrate naturally, but when they return as adults to nest, it will be to the new hacking location.

In Wisconsin, we've successfully hacked trumpeter swans, peregrine falcons, and Canada geese. We've donated bald eagle chicks to other states to help them restore their populations. John and I used to drive to Green Bay to help round up fledgling Canada geese for hacking in other parts of the state. We'd band and neck collar them, and release them in Ashland and Iron Counties. After fledging, they would return as mature two-year-olds to nest near where they'd been released. For geese we could measure our success by reading the neck collar numbers. One year the wildlife manager in Green Bay, Tom Bahti, asked if we would take some molted non-flying adults as well. He had an overpopulation of geese in the parks, golf courses, and beaches in the area where their gooey droppings created a nuisance and health hazard. We agreed to take the birds, but about three weeks later Tom called John and said, "Your geese are back." After they'd molted, the adults had regrown their flight feathers and returned to Green Bay. Unfortunately, all the young had followed them. That was the last time we took adults, having documented that, as the data supports, hacking only works with young birds.

In 1997, our Bureau of Endangered Resources got a call from the Iowa DNR asking if they could have some of our chicks to hack as they had no breeding osprey in the state. Following a time-honored tradition of interstate cooperation on matters such as this, we said sure,

When I heard about the Iowa osprey project my ears perked up, especially when I heard that two of the proposed hack sites were at Lake Macbride and the Coralville Reservoir. These lakes were boyhood fishing haunts about 20 miles from my family's farm. Bruce Bacon and I were heavily engaged in osprey work at the time. We were installing platforms, banding, and were conducting an osprey food availability study. The project used newly developed camera technology to record food deliveries and chick survival, and we had successfully used a method to capture adult osprey at nest sites using a noose carpet trap. This involved a wire mesh dome covered with small monofilament fishing line loops that would close around the foot of an unsuspecting osprey and snare it when it landed on the nest. We had caught several birds using this method and now had colored numeral bands on adult osprey as well.

With all our osprey experience, we were a little taken aback when Lowell Tesky from Hayward and Dave Evans, "Fud," from Duluth got the nod to lead hacking the osprey to Iowa instead of us. Lowell was the wildlife technician who did the eagle and osprey census surveys by airplane. Fud was a contract eagle bander who worked all over the region. These guys were friends of mine, so I swallowed my pride, realizing that there were some other good reasons they were chosen. Lowell had a larger osprey population with greater productivity in Sawyer County, and Fud, who was doing the climbing, lived a lot closer to that county as well. Of course, I couldn't resist pointing out that the reason Lowell had this abundance of osprey was because they were living on platforms John Olson and I had put up.

I was still stewing a little. I used to join Lowell on his eagle and osprey flights when he flew Iron County. If he had the right weather, he'd call me at 5:00 a.m. when he left the Spooner airport and he would pick me up at Manitowish Waters' airport. In the early days of osprey recovery, he'd do two flights per year per species. The first flight was to record nesting and the second for chick productivity. It was really neat to look down and count the nestlings from the plane, but it was hard on me, as the circling in the small plane tended to bring on air sickness and I would end up sick nearly every flight. Lowell had an iron stomach and never got sick.

Lowell consulted with me on their hacking plan and I suggested one change. The draft protocol was to take one chick from each nest with two or more chicks. Iowa was paying good money for the birds, so the idea was to pick the largest, healthiest chick. I told Lowell to do just the opposite, to take the smallest chick if he wanted Iowa to be more successful in establishing a breeding population. Like most birds of prey, ospreys have reversed sexual dimorphism, which means the females are larger than the males. An adult osprey female may weigh four pounds, males around two and a half. If they were to take the smallest nestling they would most likely take more males, which I suggested would yield better results.

After osprey chicks fledge they migrate south to the Florida Keys, across to Cuba, and then on to Central and South America. After two years, surviving birds return as mature, breeding adults, migrating to near their natal site to nest. Studies have shown the male takes the female to his home, not the opposite. For example, if an Iowa female mates up with a Michigan male, there are greater odds that the pair will move to Michigan, his home, not her home in Iowa. If we selected more males for Iowa, the chances would be greater for mated pairs to return and nest near their hacking sites at Lake Macbride and Coralville.

There was one other silly thing that bugged me about this hacking project. When I was a teenager back in Iowa, our family subscribed to a monthly publication called the Iowa Conservationist, published by the Iowa DNR. When it arrived in the mail, I immediately read it cover to cover, devouring the scientific articles about Iowa wildlife. Sometime in the 1960s, Iowa hired a furbearer biologist out of the University of Iowa named Ron Andrews. Ron's master's thesis was on red fox. I was about 15 at the time and read every red fox article Ron wrote. I was learning to trap fox back then and wanted to know everything about their biology and natural history. I had never met Ron but considered him a mentor and followed his career. I knew that besides furbearers, he was in charge of endangered resources for the Iowa DNR.

In 1997, the Midwest Fish and Wildlife Conference was held in Madison, Wisconsin, so Bruce and I were able to attend. Bruce introduced me to a friend of his, a woman named Jamie Edwards, who had once worked part-time for the Wisconsin DNR and now worked for the Iowa DNR. I told Jamie I was originally from Iowa and, during our conversation, asked her what projects she was working on. She told me that, among other things, she was in charge of the osprey hacking project.

That got my attention and the gears in my head started turning. Bruce and I described our platform and banding work and I told Jamie that my family farm was only 20 miles from two of her hack release sites. "Jeff," Jamie asked, "Why don't you come down sometime and do a workshop about osprey management, platforms and banding for the Iowa DNR employees? We're getting your osprey but we're new to this work. I'll have to check with my supervisor but if you're willing, I'll try to arrange it." I asked who her supervisor was. "Ron Andrews," she said.

That did it, I went into action. I knew working the bureaucratic channels would take some finesse. It would be unusual for a field tech to lead an out-of-state project but I found an ally from Endangered Resources, my old friend Randy Jurewicz. I laid out the plan and told Randy I'd make it a family visit, drive my own car, and not charge for time or mileage. Randy said, "Great, do it. But put your time down. I've got some money in the budget." I was set.

The day finally came. At the time, I was dating a new girlfriend, fellow biologist and my future wife, Terry. We had driven down the day before and visited my family before meeting Jamie at the Coralville DNR building. She gave us a quick tour of their hacking tower. It was designed to resemble a nest looking out over the water. It was predator-proof and allowed fish to be placed inside to feed the young without the osprey seeing the feeding staff, an important step in preventing the young birds from imprinting on humans.

When it was time for my talk, the conference room was packed. Jamie had invited every biologist working for the Iowa DNR to the session, as well as the county conservation boards that were hacking osprey in their areas. I gave our slide presentation, explained our banding program, and showed them our banding gear. I brought in an osprey platform, discussed how they were designed and built, and answered questions the rest of the morning. After lunch, the attendees became my ground crew while I topped a tree and erected a platform. Later, Jamie told me her boss, Ron Andrews, wanted me to ride with him and pick out some platform sites in the neighboring Lake Macbride area. As you might imagine, we spent hours talking about his red fox research.

Loon Research Days

"Above came a swift whisper of wings, and as the loons saw us they called wildly in alarm, increased the speed of their flight, and took their laughing with them into the gathering dusk. Then came the answer we had been waiting for, and the shores echoed and re-echoed until they seemed to throb with the music. This was the symbol of the lake country, the sound that more than any other typifies the rocks and waters and forests of the wilderness."

—Sigurd F. Olson, *Listening Point*

Barefoot Jerry

If you have spent any time in loon country, you will know that people love their loons. Whether it's in Maine, New Hampshire, Minnesota or Wisconsin, loons are popular symbols of the northwoods. One day in 1985, Ray Anderson stopped by our office to talk over a loon project. Ray said he'd hired an assistant, an undergraduate, and was sending him north for his first field season. We'd help him all we could, wouldn't we?

A few days later, in walked Jerry Belant, a.k.a. "Barefoot" Jerry. He'd grown up in Merrill, Wisconsin, about an hour and a half south of Mercer. His dad had been a mechanic for the Chevrolet dealership down there. Jerry drove an old Chevy and didn't know much about loons, but neither did we. He told us his project objectives were to catch, band, and color-mark enough loons to establish some baseline data on their natural history along with testing radio-telemetry techniques. Paul Strong, the Project Loon Watch coordinator at Northland had piloted a color-marking system with specially formed bird bands made out of stiff vinyl plastic of various colors that would conform to a loon's flat leg. When these colored bands were placed in specific combinations, each bird could be individually identified. Once you could identify a specific bird, the whole world of loon research opened up to you. You could document natural history characteristics such as nesting territory, home range size, social behavior, and parental roles. Scientists were interested in these, and other mysteries—such as whether loons mated for life or switched mates each year, did they return each year to the same lakes, which bird arrived first—the list went on and on. Jerry explained that part of his job was to figure out a way to catch the birds, get them in hand, so they could be banded. A search of the published literature suggested that no one had found an easy method.

I had caught and banded hundreds of birds—ducks, geese, swans, ospreys, eagles, merlin falcons, owls, and woodcocks—so I thought catching loons would be easy.

"No problem," I told Jerry, "I'm your guy."

Jerry ended up renting a little house on Mercer Lake Circle that we dubbed the "loon research station." Because Jerry had almost no money for expenses, Jack Kunath, the game warden, took pity and gave Jerry's team "blanket authorization" to obtain and eat any roadkill they scavenged. One day, Jerry told us about a particularly choice specimen that was bloated and off-gassed while gutting, causing a crew member to lose his breakfast. Nonetheless, we found them some temporary freezer space in the DNR wildlife freezer among our scientific specimens. A trip to the Rowe's potato farm an hour north to Gurney got Jerry 100 pounds of potatoes for $4.00. He then loaded up on macaroni-and-cheese dinners, canned tuna, and a bag of onions, and his larder was complete. I was impressed; Jerry was getting organized.

A few days later another student arrived to join the loon crew. His name was Matt Giese, and Ray Anderson had hired him as a work-study student at UWSP. He would be Jerry's assistant. He drove up from his home in Chicago, arriving in a Dodge pickup that had seen better days but seemed reliable enough for field-work. Jerry and Matt got started by reviewing the published literature, looking at the work Paul Strong had done capturing loons in Maine. They also reviewed papers by Dr. Judy McIntyre (Utica College), at the time one of the foremost loon researchers in the country, who had caught a few loons in upstate New York. The largest number of loons had been caught in Minnesota by a duck banding crew. These incidental loon catches had all been accomplished using a high-speed chase boat at night with a musky net and high intensity spotlight. This was the same technique I had used to catch and band hundreds of mallard ducks.

Jerry formed a crew and tried the high-speed capture technique with little success. The loons wised up fast, were swift and far more elusive than ducks. We realized that ducks' and geese's escape mechanism is to fly up and off the water when threatened, but in summer they molt and have a flightless period which a capture crew can take advantage of. During this time, you could just motor up, mesmerize them with your light, and scoop the net down over their heads. Loons were quite the opposite. Instead of flying up to escape, they quickly dove and you had to scoop your net down under them from the front. Picture the process: the boat is running at high speed with the net-person kneeling in the bow. As soon as the boat is next to the loon, the netter plunges the net down, and then holds on for all he was worth as the drag from the water nearly yanks the net out of their hand, or they are pulled overboard.

Jerry and Matt surveyed many of the lakes around Mercer and found plenty of loon pairs. Each night we would set out and try again. We would wait until about 10 p.m. for it to get fully dark and usually got two attempts per lake per night. We quit around four a.m. when the sky started to lighten. I didn't go along every night, as I still had my other DNR duties and this was extra "volunteer" time. Night after night we were defeated, exhausted, and discouraged.

Let's Take My Truck

One night, Jerry and Matt asked me to come along for another try. Paul Strong had told them loons were easier to catch when they had chicks. They didn't dive as much then because they want to stay above water to defend the chicks. Jerry and Matt had found some newly hatched chicks on Duck Lake, a small 40-acre lake in the middle of the county forest. It was an hour's drive down a rutted road and it had no boat landing, so we would have to drag their 12-foot boat 200 yards through the woods in order to launch. They also had to haul a heavy car battery, which they used to power a million-candle-power light. Then there was the outboard motor, a gas tank, and their capture gear to portage. You can see why they needed me.

When we were loading up at the ranger station, I took a good look at Matt's dilapidated truck. "Are you sure you don't want me to drive?" I said, "Let's take my truck." They said no, assuring me we would be fine, and besides, we were already all loaded up.

"Hop in, relax, take the window seat," Matt said.

"I'll take the middle," I said. "If a wheel falls off this jalopy and we crash, at least I'll have you two for padding."

We got to Duck Lake, hauled in our gear, and portaged the boat. We could see the loons out on the lake. It was starting to get dark. Matt lifted the truck hood, and when I asked what was wrong, he said we needed the truck battery because their spotlight battery had died three days ago, and the truck battery worked fine. The truck alternator would charge it right back up.

When it was dark enough, we headed out. Matt drove the boat while Jerry handled the net. I stood in the middle and held the spotlight and cord. We circled the lake, shining our light and looking for the reflection of our beam on the loon's eye or the white flash of plumage. We spotted the family and turned toward them at a good clip. One adult dove immediately, but the other hesitated. It let out a tremolo alarm call near the chicks, Jerry scooped hard and fast, and when he pulled the net up there was a loon! We didn't have time to celebrate because we now had to figure out how to handle a loon in the hand. I soon discovered loons would be the most challenging birds I ever worked with. Their bills were ice-pick sharp and they never quit fighting. Ducks, geese, and swans calm down quickly after capture and were a piece of cake compared to these feisty birds.

Once we had the bird in the boat, we followed the banding protocols Paul Strong had set up. The right leg got the aluminum U.S. Fish and Wildlife Service band on the bottom with a yellow plastic band on the top. The left leg got a unique color combination, green over red. We didn't want any other loon in North America to have the same color combination. This was important because we wanted to re-observe birds year after year and needed to be able to see the color leg band combination. Because loons keep their legs underwater most of the time it is not easy to see color bands. The most likely times to observe bands are when they preen their feathers. Loons have an oil gland at the base of their tail that they use to waterproof their plumage. When they preen, they rub their bill with the oil and then coat their feathers. Preening occurs about every two hours, and as they preen their sides and belly, they roll halfway over, tipping their leg and foot out of the water. When they preen the other side, the opposite leg is exposed for viewing. Researchers call this a "roll preen." Loons also have an odd behavior where they stick a leg out of the water and shake it in the air. This is called a "foot waggle," and while no one is sure why they do this odd behavior, one theory is that it assists in thermo-regulation. Their feet have a large surface area and the evaporation off the wet skin could be cooling; they also tend to foot waggle more in warm weather.

After we had the bird banded, we had additional information to collect and Jerry and Matt carefully recorded physical data including weight, length, bill length, and foot measurements. They clipped part of a wing feather for mercury analysis and took a blood sample for DNA and chemical analysis (toxicology). At that time, the only way we knew we could determine a loon's sex was by testing its blood. Years later, after capture techniques had been refined and we had data on a large number of birds, we learned how to sex loons by weight, having learned that females are generally about two pounds lighter than males, which can weigh up to 10 pounds.

Jerry and Matt finished up by recording the capture and handling times and we released the loon in good condition. We were ecstatic as we portaged the boat and all our gear back and loaded Matt's truck for the return trip home. When Matt turned the key all we heard was a series of clicks. We'd used up the battery while processing the loon. The starter would not engage, and the truck was an automatic so we couldn't push it and pop the clutch to start it as you could with a manual transmission. We were stuck, and we realized with some dismay that we hadn't told anyone where we were going.

We got out the map. I knew the nearest lake, Island Lake, had some year-round residents and phone service. It was a longer route home but the shortest distance to a phone, so we started walking. We headed out about midnight, and then it started to rain. I know it sounds bad, but we actually had a jolly time on that ten-mile walk. We got to Island Lake about five a.m., just as it was starting to get light. There was a phone line coming into the third cabin we came to, so we tried the knob to see if it was open. It was locked, so on we walked. A few minutes later a truck came down the road and pulled up behind us.

"What are you fellas doing out here?" a man asked. We explained our predicament. "I get up real early and I was having a cup of coffee when I saw your flashlights. I live on the other side of the lake. I thought you were burglarizing my brother's cabin. I've got his key right here," he said. "Sure, you can use his phone."

Now, who do you call at five a.m., and whose number can you remember off the top of your head? We chose Tom Cawst, who worked for fish management. We thought he was single and would understand. Tom came and got us and brought jumper cables. In the future when I joined the loon capture crew, I insisted we take my truck.

The Loon Trap

Jerry Belant's loon study was moving along. The loon capture component wasn't going very well, but Jerry was collecting a lot of other important loon data. He was learning about nesting and hatching dates, nest habitat preference, loon population densities, nest predation and nest habitat characteristics. But he still needed to catch and band more birds because he needed a large sample size to make definitive conclusions.

One day when Ray Anderson was up from Stevens Point meeting with us, John Olson had an idea. The previous year, John had worked with Rick Newton, another of Ray's students. Rick and John had floated a loon decoy, stretched a fish gill net around it, and propped up on wooden stakes in shallow water. They'd played a series of loon calls from a tape recorder, and a loon had responded and had swam over to the decoy, attacking it and ripping right through the net. They hadn't caught the bird but were amazed how aggressively it had reacted to the calls and the decoy.

John suggested we try that technique again, and after some discussion Ray agreed. "Wait a minute," I said. "I've got an idea." I ran to the ranger station shop and grabbed a Bailey live catch beaver trap. I set it and sprung it to demonstrate how it worked. The trigger is in the middle, and when it is bumped the jaws slam shut capturing the beaver in a chain-link mesh wire net. I suggested, why not attach a floating loon decoy to the trigger and place it in about six inches of water within a loon territory? Then we could play the defensive loon calls used on Rick Newton's experiment, and when the aggressive territorial loon comes to attack the decoy, it would set off the trigger, closing the jaws with the loon inside.

They didn't like it. It wasn't big enough, they said. You couldn't set it in deep water and no loon would go into shallow enough water to set off the trap. If we injured a loon we'd be in big trouble. When John and Ray left, I started trying to convince Jerry and Matt. We could make this work, I said. We could attach blocks to the jaws to make them offset. That way we wouldn't hurt a loon if a neck or foot were inside the jaws. We could replace the chain mesh links with soft nylon net. We could make this trap twice as big by building a larger frame and larger jaws. Then we could set it deeper, maybe even float it somehow. These traps are manufactured only 50 miles away in Tomahawk, Wisconsin. Why not go down and talk to the company's engineers? Maybe they could help.

The next day Jerry and Matt headed for Tomahawk. The people at Tomahawk Live Trap Company understood what we needed and provided advice. Back at the DNR shop Jerry and Matt modified the jaws and springs. All we had to do was take apart one of my beaver traps for the trigger assembly and find some nylon mesh fish net. The DNR fish technicians had net we could use. Next Jerry found a duck decoy and painted it to look sort of like a loon.

A few days later Ray Anderson stopped by on his way to the Apostle Islands where he had been checking up on his eagle and bear researchers. Yes, he said, it could work. It might be worth a try. "But that's the ugliest loon decoy I've ever seen." A week later he brought the loon tape recorder and a hand-held megaphone. He also handed Jerry a beautiful hand-carved wooden life-size loon decoy. It was a perfectly painted coffee-table-quality model.

On our first try we were at Trude Lake on the Turtle Flambeau Flowage. Jerry was hidden on shore, megaphone in hand. Loons have four main calls. The "hoot" is a call for non-aggressive close distance communication. It's like a loon saying, "How are you doing, how's it going?" The "tremolo" is a frantic call, which expresses distress and danger, for instance, if a snapping turtle is after the chicks, or a boater is too near the nest. The third call, the "wail," is used for long distance communication. It's equivalent to saying, "I'm over here. I'll be home late for dinner." Then, finally, there is the "yodel." Only the male loon makes the yodel call, and it is definitely territorial. It means, "I'm the dude here. Stay out unless you're itching for a fight."

I know all this now, after 20 years of loon research, but back in 1984 we didn't know anything about the meanings of loon calls, so we played them all. We placed the decoy floating in the middle of the trap attached to the trigger. The loon heard the calls, it saw the decoy intruder loon, and attacked. We saw the loon submerge about 20 feet in front of the decoy. We saw the V in the water as it submarined toward the trap. It hit the decoy, the trap sprung, the jaws rose and closed—and the loon surfaced about 30 feet from the trap and paddled away.

We went back to the drawing board. Jerry replaced the netting with larger, lighter mesh to speed up closure time. Next, he attached a long, clear fishing line to the trigger so he could control when the jaws would close by pulling the line with his hand. Lastly, the beautiful hardwood decoy was replaced with a soft styrofoam decoy. We'd found a sliver of the loon's upper bill stuck in the hardwood. Don't worry, it was only a tiny tip. They grow back like fingernails. But we weren't going to take a chance again on injuring a loon's bill.

The next try Jerry caught a loon. He caught three with this method, but it wasn't perfect. Once in a while he misjudged when to pull the line. As the season progressed, we discovered the method's main drawback was it only worked in the pre-nesting season, when the loons were hormonally charged up and would aggressively defend their territory from intruding loons looking to usurp their nesting area. If we set the trap up after the chicks had hatched, the adults just moved away to the other side of the lake. They took the chicks with them and stayed away from us and our decoy.

Close, But No Cigar

At county fairs, the High Striker strength game challenges players to hit a pad with a mallet, sending a ball up a tube to ring a bell. Historically, a common prize was a cigar, but most of the time the ball falls just short, leaving the player prizeless. This is what happened to us one night when we unexpectedly caught an adult loon but totally missed a major research breakthrough.

That night, Jerry, Matt, and I were on the Turtle Flambeau Flowage. The loon pair we were targeting had one chick and we had tried for over half an hour to catch either adult. We came close once, and almost had an adult in the net, but it dove and got away. When you come that close you tend to educate the loon, and we decided our chances on that pair were probably over for the night.

As a consolation, we decided to go for the chick. Jerry had some chick web tags that he attached through the foot. Chicks were easy to catch when young because for the first few weeks they didn't

dive very well. We easily caught this one and turned off the boat motor, focusing our lights and attention on banding the chick. I was gently holding the chick and it was vocalizing a typical chick distress call, a high-pitched peeping they often make when captured or stressed. Suddenly, right beside the boat we heard a loon tremolo call and Jerry dropped what he was doing and grabbed the net. The chick started vocalizing more, and the adult called again and approached the boat, staying on the surface. Jerry dipped the net down and easily scooped it up. We banded and processed both birds and then released them together.

Five years later, a copy of the *Tremolo*, the quarterly newsletter put out by LoonWatch, arrived in my mailbox. An article, written by my future wife Terry, told about a new loon catching technique perfected by David Evers, who was working at the Whitefish Point Bird Observatory in Paradise, Michigan. Evers had been doing loon work at Seeney National Fish and Wildlife Refuge in Michigan's Upper Peninsula and had made a giant leap forward in the loon capture world. His method was to locate a pair of adults with chicks, then after dark three people in a canoe or small boat would slowly approach the loon family and mesmerize them with a spotlight. They would first play the loon wail call and when they got close, they would change to a chick distress call. When the adult loons heard the distress call, they would allow the boat or canoe to approach them. Sometimes they even swam toward the boat and were easily scooped up in a musky net. Dave Evers had caught over 100 adult loons using this technique and I was flabbergasted. We had been so close to this discovery on the night I'd held that chick. The chick had called and the adult approached, allowing us to make an easy catch, but we just hadn't connected the dots. Today, this capture technique is the standard and using it, thousands of loons have been banded from Wisconsin to Alaska. That night on the Flambeau, Jerry, Matt and I were close, but no cigar.

Loon Research Station #2

After that first trial year, Ray Anderson was quite impressed with Jerry Belant and asked if he'd like to be his graduate student and do a master's thesis on loons. Jerry said yes, and this meant he was coming back to Mercer for two more years of loon research, but with a few study modifications informed by our first year.

Most importantly, the study site would focus on the Turtle Flambeau Flowage. During Jerry's pilot year he had discovered that the TFF held the mother lode of loons in Wisconsin. The 14,000 acres of water include perfect loon habitat, 114 miles of mostly undeveloped shoreline and 195 islands. At that time, it was owned by Northern States Power Company (NSP). Today it's owned by the DNR, purchased with money from the Knowles Nelson Stewardship Fund. This was a loon researcher's dream, as 30 nesting pairs could be accessed from one boat using one landing. Working with a sample size this large on smaller inland lakes, which often have one pair of loons per lake, would have taken more travel time and funding. There was one other bonus: I had just purchased 40 acres of land on an island right in the middle of the TFF where I was planning to build a cabin. We were soon to have Loon Research Station #2.

As a kid growing up in Iowa, I'd always dreamed of living in a log cabin in the woods. When I worked for the DNR out of Park Falls we had an end-of-summer cookout party on an island in the TFF each year. At my first cookout it had rained briefly, and then a rainbow came out and the sky cleared. I remember the sensation of standing there looking at that rainbow from a high point on the island. I'd been in the Boundary Waters Canoe Area Wilderness in Minnesota as a Boy Scout when I was 15 years old. Looking out over the TFF reminded me of that experience.

After that party I started looking for land. I wanted 40 acres, but I didn't want just any 40; I wanted it to be a special place. I kept an eye on real estate boards and looked through plat books. I wanted to be touching public land so development couldn't move in next to me. What impressed me about northern Wisconsin was that unlike Iowa, Wisconsin had vast amounts of public land. Wisconsin's abundance of national, state, and county forests provided miles of land for hunting, fishing, trapping, berry picking, mushroom hunting, camping, and just hiking and exploring. The northwoods offered a freedom to wander that Iowa lacked.

One day two foresters from Owens-Illinois Paper Company were at the ranger station. Dave and Steve stopped by regularly to borrow aerial photos to study some of the timber tracts they managed. I asked them if they could spare a minute, I had a question. I opened up the plat book and pointed out a 40-acre parcel their company owned on an island in the TFF. "Oh yeah, that parcel," they said. "We know what you're up to. You want to build a cabin, don't you?"

"Is there any way I could?"

"Okay, we'll give you the scoop," they said. They told me their company had bought a large tract of land many years ago in one big real estate deal. There were also a few scattered 40s thrown into the mix that Owens hadn't really wanted, but the deal was "all or nothing," which was how they'd ended up with the 40 acres on the TFF island. They said that Owens would probably never log it as the nearest road was three miles away. It would require the loggers to freeze down a road across the ice to get there, which would be too costly for only 40 acres of timber. Owens' company policy was to never sell land because the company had a timber base to maintain to feed their paper mills. Dave and Steve ended their explanation stating that in the past the company had traded land, but only if it was to their advantage. I gave them my thanks and they left.

That night I started paging through the plat book again, looking for a possible parcel to trade. After a few minutes I found it, a 40-acre parcel listed under the ownership of Gus Ferek. It neatly connected two parcels of Owens-Illinois land, an 80-acre tract and a 160. The next time Dave and Steve stopped I showed it to them. "Yes," Dave said. "We've just been up that way. We have a timber sale starting soon in that area and a trade might work." To make a long story short, it did. Gus Ferek's land was for sale. I had to throw in an extra $2,000 but closed the trade for 7K.

Now that Jerry was moving his loon study to the TFF, I jokingly offered him my future cabin for his loon research station, saying it would save budget if he didn't have to pay rent. The only problem was I didn't have a cabin—we'd have to build it first. So that summer, Jerry and his crew helped me clear a cabin site.

I wanted to build a Scandinavian-style scribed cabin where the logs are carved to fit tightly without chinking. You mark the log on top with a tool called a log scribe then roll it back and cut or chisel out the lateral groove to fit the contour of the log below it. The ends are saddle-notched so they fit tightly together. It's not easy to build this way, taking way more skill than I had. In Minnesota there are log building schools that teach this technique, and I was lucky and found just the guy to help me—or more like, let me help him. His name was Alan Sabatke, or Ski for short; he'd been to log school, and he'd already built a few cabins. Our deal was that he would build the cabin for me in my yard at Mercer for $3,000 labor. I would buy the logs and I do all the grunt work such as peeling the bark off the logs and grinding the notches. He would also help me disassemble and move the logs to the island.

Twenty miles away, in the town of Springstead, a logger named Dick Cramer had a nice pile of red pine he'd salvaged from tornado damage. Ski looked them over and I bought 60 logs for $2,500. Ski said that, allowing for the overhang on the ends past the notches, he thought we could build a cabin about 20x25 feet in size. "Ok," I said, "let's get started."

At the end of the first day, Ski said, "I usually live in the basements of my job sites. I'm not renting anywhere right now. Would you mind if I slept in the back of my van or pitched a tent in your yard?"

"I do mind," I said. "I'm not letting you sleep in my yard. Come on inside. My couch has a pull-out bed." So, Ski moved in with my son Zach and me and we completed the log structure part of the cabin in seven weeks. Half the town stopped to see what we were building in my driveway. Jerry and Matt helped peel logs and slide them up our ramps as the walls got higher. Friends from Iowa also came to help. One friend, Mark Jaunzems, a seasonal botanist with the DNR in Madison, came for three weeks. One day when we were on the fourth course of logs, a man from the Wisconsin Department of Transportation stopped by. He started measuring how far the cabin was from State Highway 51, and wouldn't you know it, we were seven feet too close. Ski told him it wasn't a permanent building and we'd be moving it as soon as it was done. "It doesn't matter," the DOT man said. "It's illegal to build that close to the highway."

A few days later I received a letter stating that we had 10 days to move the cabin or a citation would be issued. There was no question of ignoring the letter. I was a state employee and knew I should not be in violation of another state agency, so I started calling friends. They all showed up on a Saturday as planned, and out came three pairs of log tongs. A Scandinavian scribe log cabin has no bolts, spikes, or nails, all the logs fit together like a puzzle. Luckily, we were only up four courses. We took the cabin apart and reassembled it on the other side of the yard in less than two hours.

By late September, the cabin was almost complete, and Ski had become family. The walls were up, the roof purlins were in place, as well as the king post trusses on the gable ends. "Now comes the important part," Ski said, "the peeling and placement of the ridgepole." He explained that Scandinavians had a ritual to secure the good karma for the cabin. Everyone had to help peel the ridgepole and while peeling they had to chew a sliver of wood from the pole. Hardest of all, no impure thoughts could enter the crew's minds during this process. He went on to explain that after the peeling we would need to plan a ridgepole celebration. We would attach a small Christmas-tree-sized pine on the end of the ridgepole and have a big party with music and dancing.

Monica from Michigan, our wildlife assistant for a season, peeled most of the ridgepole. Ski felt she was the only one of us who could be trusted to maintain pure thoughts. And she was the only one he had allowed to help him scribe the logs. I rounded up everyone I knew who played guitars, fiddles, and banjos. We put a tarp over the roof and moved a woodstove inside. We invited everyone who had helped and about 30 people showed up. When Jerry arrived from Stevens Point, I spread my arms and told him this was next year's Loon Research Station #2. Everyone danced and celebrated and the cabin was blessed.

Now you might be wondering how we were going to get this cabin from my yard in town nine miles down the road to the Flambeau Flowage, then three more miles across water to my cabin site. Frankly, we didn't have a firm plan. We knew we could take it apart and truck it the nine miles to the boat landing but getting across the water, was the challenge. At first we thought about floating the logs, after all, in the old days, log drives had been popular and exciting. We considered building sleds and sliding them across the ice. This was also a popular method to transfer logs in the 1800s. After considerable back and forth we decided that since we only had six weeks before freeze-up sleds were our best bet.

I approached Dave Sleight, the fire control equipment operator and expert welder who'd helped design and build our bear stanchion. He said he could build me two sleds, one for the back of the log and one for the front and we could use truck chain tighteners to fasten the log securely. If fastened tight, the back sled would follow the front, which is how the old horse-loggers used to do it.

That decided, we now had to come up with a strong "horse" to pull the loads. At the time a local cross-country ski club had a double-track Skidoo snowmobile called an "Alpine" that they used for trail grooming. It was a powerhouse and the club said I could use it if I gave them a donation, so we had our work horse.

The day of the move finally came in late January. We had good ice on the Flambeau and no slush. I talked the town road crew into plowing out the road and parking lot at Fisherman's Landing boat launch. I hired a logging truck to move the three loads of logs to the landing a few days later. Ski and I had marked the end of each log, identifying its sequence in the cabin so nothing could get mixed up.

I knew there were three ways this project could be done. The first was to get just enough help with the right skills. This would mean a lot of work for those involved but the job would be done efficiently. A second option was to bring in way more help than we needed and make a party out of it, kind of like an Amish barn raising. The third way was not having enough help which would be a disaster.

Ski insisted on the second system, so I made calls and sent invitation letters to about everybody I knew. My brother Brad brought a van load of farm boys from Iowa. He also brought several gallons of Mom's chili and chicken noodle soup. The DNR crew was there too, Harold the warden and his wife Cathy, Jim and Tom from fisheries, John and Charlie from wildlife, and Marko and Dan from forestry and fire control. We probably had 30 people in all. I'd arranged housing for the out-of-towners and everyone brought sleeping bags and musical instruments.

Moving day dawned sunny and we loaded the first log by 9 a.m. and took off. Dave's sleighs worked perfectly and we were in business. In four days, those logs were moved across the three miles of ice and re-fitted together into a cabin shell. We had a crew loading logs at the landing, a second crew moving logs across the ice and a huge crew at the cabin site, grabbing the logs with their log tongs and putting them into place. Ski supervised everyone and made sure the right log came up the trail in the correct order. We placed the wood stove outside and put on pots of chili, soup, coffee, and cocoa. Everyone worked hard and had a good time and Loon Research Station #2 was in place. I thanked everyone and told them it wasn't my cabin—it was a gift from my friends, and they should feel free to come and use it anytime.

At Easter my son Zach and I visited the farm in Iowa and showed Dad the pictures of moving the cabin. He said, "I sure hope you know what you're doing building that cabin on an island. You won't have electricity, you won't have plumbing, and everything will be twice as much work. You can't even get there part of the year due to thin ice." I was silent. Then Dad said, "You might as well finish what you started."

Ski's dad, also a carpenter, had stopped by the Mercer job site once in a while. His line was, "I don't know why you guys like building with logs. Don't you know 100 years ago a machine was invented called a sawmill? It sure made construction a lot easier."

When Zach and I left Iowa, we noticed the back of the truck had filled up. There were old-time hand tools like wood files, planers, chisels, and crank drills. Mom had added cast iron pots and skillets and canning equipment. It was stuff people used in the old days before electricity and refrigeration. These tools were outdated in the modern world but would get new life at the cabin.

That spring after ice-out, Jerry arrived to start his two-year master's thesis on common loons of the Turtle Flambeau Flowage. Ski arrived too, as I had hired him to finish the rest of the cabin. By now Ski and I didn't talk money or bids or contracts, I just told him to work whenever he wanted, take off whenever he wanted, and every Friday tell me what I owed and I would write him a check. If we ran out of money we'd figure something out. "The only rule," I said, "is let's have fun building it." Ski said, "That works for me."

The logs of the cabin were assembled, but we had no door, floor, windows, loft, roof or porch so we couldn't really live in it. We needed some shelter so we opted to erect what later got dubbed "tent city." Jerry got a few canvas tents from the university and we built wooden floors to keep us off the ground. We each had our own tent, plus a tent Jerry used for an office. Of course, this research was being done the old-fashioned way. We did not have computers, just waterproof paper, pencils, and data records. Field notes were written down twice, once during the day's work and then an official record was laboriously re-written each evening and tucked away out of the elements as a backup. I had a new 16-foot boat and motor. Jerry had his university research boat, an old U. S. Fish and Wildlife Service boat with a 20-horsepower Mercury outboard that he named White Lightning. Poor old White Lightning broke down often as the TFF has numerous rock bars and stumps scattered throughout, hiding just beneath the surface of the water. These were the days before electronic depth finders, and we had no bathymetric contour maps, so we had to learn our way around the flowage the hard way, one rock, one stump at a time.

Jerry had a great crew, and over the two years had five assistants. A fun-loving guy we nicknamed Wild Bill stayed the longest. Wild Bill and Jerry never wore shoes. They maintained that hip boots were too hot and as they were constantly jumping in and out of boats going barefoot was more practical. Jerry and Bill were so popular with locals that they were even exempt from the NO SHIRT NO SHOES NO SERVICE rules at the bars in Mercer.

While Ski and I worked on the cabin, Jerry and Bill worked on loons. My son Zach lived summers with his mother in Iowa, so I stayed in tent city all I could, weekends for sure and weeknights when possible. We were all on a budget, so we threw in together on supplies. We had one standard meal, what Ski called taters and onions. Each night over the campfire we cooked a big pot of potatoes, threw in some onions and a few cans of tuna. We ate all we could that night and left the pot hanging there. In the morning we reheated it and finished it off. That, and a few cups of coffee was our breakfast. We skipped lunch—after all Jerry and Bill were out on the water anyway. We repeated the same meal again each evening. We did the same thing every day and, somehow, never got tired of it. After dinner, Jerry would bring out his banjo and we'd all sing along, and as the evening wore on our tunes went from bad to terrible. Friends would come and join us on weekends. Life was great at tent city.

When the roof and porch were finally built, we moved into hammocks on the porch. Then came the cabin floor, followed by the loft floor. By midsummer we were moved in, and in his free time, Ski became part of the loon catching crew. We used the loon trap early in the spring, then high-speed night chase later in the season. Jerry tested out some radio-telemetry packages on loons. He mapped loon territories and established the groundwork for future loon research on the flowage. The Turtle Flambeau loons are probably one of the most studied loon populations in North America, and Jerry's master's thesis sits on my bookshelf at the cabin.

Loon Research Station #2 eventually became, simply, the cabin, but researchers continued to come. Next to Jerry's thesis on that bookshelf there are several other theses and doctoral dissertations. Jim Peruk came from Idaho and did his doctorate on the Flambeau loons. Mat Selinski from UW–Eau Claire did a master's on merlin falcons. Jerry Hartigan did master's work on loons. Cheryl

Dykstra did her doctorate on the Flambeau's bald eagles and the impacts of mercury. Keith Warnke did his master's on bald eagles. Dr. Mike Meyer conducted over 20 years of mercury studies that included the flowage loons. Dr. Jack Stewart of Northland College brought his students down five different years to study social interaction between bald eagles and ospreys. Kevin Kenow from U.S. Geological Survey (USGS) studied Turtle Flambeau loons. Tim Lewis from Allston Ohio State University brought his students to help band ospreys for three years. The Iowa DNR sent personnel to learn osprey banding after we reintroduced the species to Iowa. Earth Watch crews came to band loons. Dave Evers and his crew from Biodiversity in Maine came to band loons. My wife Terry and I taught a wildlife management course at Northland College one spring and brought the students to the cabin for two weeks of field study. In 2012 another Northland student, Steve Thielking, did a camera study of loon nests to determine predation and productivity rates.

Today, Terry and I live at the cabin as much as we can. We cross by boat or canoe in the summer or snowmobile or skis in the winter. We have a hand pump for water, a woodshed, a sauna, an outhouse, a root cellar and a guest cabin. Grandkids come over often. We've modernized too. A solar cell on the roof powers five LED 12-volt lights. We have a solar refrigerator with an ice box. We eat better too. When guests stop by, we never serve potatoes, onions, and tuna.

It's in Their Blood

"One man alone can be pretty dumb sometimes, but for real bona fide stupidity, there ain't nothin' can beat teamwork."

—Edward Abbey

Dr. Mike Arrives

I used to believe wholeheartedly in this Ed Abbey quote. That all changed when I started working with Dr. Mike Meyer, a wildlife toxicologist. Mike is a great field biologist, a trait that I found rare among PhD folks who seem to spend more time behind a desk than in the field. He grew up in Eagle River, an hour east of Mercer. After grad school, he came back to Wisconsin because he loved the lake country.

Mike was the first wildlife toxicologist the DNR hired, and he worked in the Bureau of Research, first in Madison and later out of Rhinelander. Mike was concerned about the impacts of mercury on wildlife and humans. Testing showed some of our lakes had fish with high levels of mercury in them, some of which are consumed by people. Mercury damages the nervous systems of wildlife and humans. Many northern Wisconsin lakes have fish advisories warning children and women of childbearing age to restrict the number of larger gamefish such as walleye, pike, or bass to one meal per month. This was and is particularly alarming for Native American families who have traditionally eaten lots of walleye and other gamefish. It also upsets some anglers who wonder why they should work so hard to catch a fish if there are advisories against eating them.

Animals at the top of the food chain (apex species) bio-accumulate (store) mercury and other chemicals in their tissues. Mike was interested in studying loons and other wildlife as bio-indicators for toxins like mercury. If these species were impacted by mercury, they could serve as a surrogate for measuring the health of an entire aquatic system. Eagles, ospreys, and loons fit this category because they eat a lot of fish, some containing heavy mercury loads. Years earlier, Jerry Belant had caught a loon near Mercer at North Bass Lake. Tissue samples he sent to the state diagnostic lab for testing registered extremely high mercury levels.

Mike had met with the folks at LoonWatch as well as field staff to consider study areas and ways to collaborate. Loons are a good species to use for mercury studies because they bio-accumulate mercury in their blood and feathers, but unlike eagles and ospreys, loons have a characteristic that makes them ideal for studying individual lakes. Eagles and osprey fish from a variety of locations and feed their young high in treetop nests; loons only feed their young prey items they capture on the natal lake. In addition, loons don't transport food while flying, nor do they regurgitate food from their crops into the mouths of their young.

Because loon chicks are only fed prey from an individual lake, blood and feather samples reflect the chemistry of that waterbody. If Mike could collect blood and feather samples from loon chicks before they fledge, he would have mercury data exclusively for that body of water. Mike picked out study areas focusing on Iron, Vilas, and Oneida Counties where lakes could be selected with water chemistry for high, medium, and low pH. He was looking to see if there was a correlation between lake acidity and mercury with related impacts on the loons. We were back in the loon-catching business.

Fud and the Eagles (A Digression)

Mike Meyer wasn't just interested in loons and mercury. Karin Kozie's master's thesis from 1985 showed that eagles on the Apostle Islands had high levels of the toxins PCB and DDE in tissue samples. The population of eagles on the Apostles at that time was small, only three pairs, and appeared to have higher turnover rate and lower chick success than inland eagles. Mike wanted to explore what factors might contribute to this difference and the National Park Service was interested as well.

Mike's first eagle researchers were Cheryl Dykstra and Keith Warnke, whose work included blood-testing bald eagles for toxins typically found in Great Lakes fish. This project was Cheryl's PhD, and although Keith had been hired as a research assistant, part of the data later became his master's thesis. Their field work started in 1990, and focused on comparing toxins and chick production of eagles in the Apostle Islands National Lakeshore and Lake Superior shoreline to those nesting on inland lakes. To do this work, they would need a control population of eagles for comparison, so Mike selected nests in Vilas, Oneida, and Iron Counties, where nest success was much higher and fish contaminant loads were lower. An added benefit was that Iron and Ashland Counties bordered Lake Superior, and we had some eagles nesting on the shoreline.

I got involved in the eagle work because I worked in Ashland and Iron Counties, and I could climb trees. My experience with osprey platforms and banding came in handy. Dave Evans, aka Fud, was Mike's main climber, but I filled in if he wasn't available. Back when bald eagles were on the federal and state endangered species list, Fud, Chuck Sindelar, and Ron Eckstein from Rhinelander DNR used to band the entire state. Fud was thin and wiry. Ron Eckstein was tall and lanky. They had the two body forms that made the best tree climbers. Fud was legendary in the climbing community, often climbing by simply holding on to the bark with his fingers.

The second year of Keith and Cheryl's eagle study involved nest cameras that would allow us to better document behavior between the inland eagles and the Lake Superior population.

Julie Van Stappen, a biologist with the Park Service, helped me put up the first of Mike's cameras, working during winter before the eagles returned. The nest was near my cabin in an old aspen and resembled an inverted umbrella. It was a very challenging climb; I had to throw a rope over the top, which Julie held tight from the ground so I could crawl up and over and into the nest bowl. We ran a cable wire down the tree, through the woods and attached it to a battery box. Once a week, Cheryl or Keith would replace the battery with a charged one. I did the climbing when something went wrong such as a squirrel chewing through the cable or a camera out of focus, but Fud did most of the treetop work.

One day, Keith and Cheryl and I were putting up a weather station on the TFF. They wanted to compare weather experienced by the two eagle populations. We picked out a tall white pine near some of our eagle camera nests.

Cheryl and Keith asked, "Can you give us some tips on climbing? We bought some climbing gear. We've hinted to Fud that we'd like to learn but he hasn't offered."

"Of course, he hasn't," I said. "You probably think it's because it's job security for him and he doesn't want competition. But more likely, he's worried about you and wants to keep you alive." Fud once told me he'd fallen twice in his career. Once from a height of 40 feet where he landed in soft forest duff and bounced up, walking away with just a couple of cracked ribs. Another time he fell only 12 feet but broke his back. I had no such compunctions, appreciating their zest for adventure, and showed them the basics and they both climbed the weather station tree.

Both Keith and Cheryl became great climbers. Cheryl was made for the job. In tree climbing terms, it's a great asset to have long limbs for reaching the sparsely placed branches, and to be lightweight so you can climb on less sturdy or dead branches with less chance of them breaking. Cheryl was also strong, which is really important. She picked it up fast, and after she completed her doctorate and moved to Ohio, she would come back each July to help us band osprey. One time, she climbed up to an osprey nest just when an adult was returning to the nest with a fish. The adult landed when Cheryl was just below the nest and didn't see her. Its tail feathers stuck over the nest, so Cheryl reached up and grabbed it. I couldn't believe it. We had banded hundreds of osprey but never caught an adult like that. Somehow, she got that adult osprey bagged and lowered to the ground without getting impaled by a talon.

The only problem I ever had with Cheryl involved her long hair. I had taught her to rappel down with a figure-eight descender. One day, Cheryl was rappelling out of an osprey tree when she went past some limbs that lifted her hair and fed it into the descender. She was stuck in midair with this wad of hair wound around the descender. She couldn't go up or down and was in great pain. I had to climb up to her and take my knife and cut out a chunk of her hair. After that, she always wore a tight-fitting hat. Cheryl now lives in Ohio and conducts a long-term red-shouldered hawk project. She's climbed and banded hundreds hawks.

One year Cheryl brought her hawk research assistant, Jeff Hays, to help us with the osprey banding. Jeff fit the credentials of a raptor gaboon to the nth degree. All he talked about was birds of prey, catching birds, climbing trees to catch birds, and equipment for catching birds. We hit it off, and he returned for many years to climb and band ospreys, eagles, and merlin. He once caught an adult osprey at the nest with Cheryl's quick grab technique. Another time he was "taloned" by an adult defending the nest when he was banding the chicks, a rare honor among bird banders. When we were on the porch that evening recounting the day's adventures, he seemed quite proud and not at all concerned about his shirt, which was bloody from the gash the talon had sliced into his scalp.

My last eagle work with Mike involved a tree on York Island up in the Apostle Islands National Lakeshore. Mike called me and said he had to take down the camera and Fud was unavailable. It was a tall tree, Mike explained, so come prepared. When we got there I couldn't believe it. The tree was the biggest red pine I had ever seen, at least 50 feet to the first limb and too big around to work a flip belt. The bark was platy and loose. I was prepared for this and had brought along a bow and arrow to shoot a light line over a limb and pull my rope up and over so Mike could hold me from the ground as I climbed. Mike and I tried shooting a line over that tree for at least a half an hour before I gave up. I was mad by then and just jabbed in my spikes and went up. I told Mike to call Fud the next time he had one like this.

Karin, Cheryl, and Keith's eagle research became a baseline data set for bald eagle toxicology in the Apostle Islands. After their studies concluded, park biologists, Bill Route and Julie Van Stappen continued to survey and collect toxicology data for several decades, turning the collective work into a long-term study. In 2017, the Apostle Islands population count was 46 nests, a tremendous increase from the three nests Karin Kozie had found in 1983. Our statewide eagle population has made a remarkable recovery as well. In 1973 we had 107 nests. Our 2022 survey found 1,700 nests.

The Loon Lady

In 1991, Mike expanded his research program to include loons, and with the loons came the loon lady. The loon lady was Terry Daulton, the LoonWatch coordinator with the Sigurd Olson Environmental Institute, who'd written that loon capture article in the *Tremolo*. Mike needed to know what lakes had loons, and LoonWatch was the go-to source for loon data in Wisconsin.

Following the first Wisconsin loon population census in 1976 by Ray Anderson's student, Gary Zimmer, LoonWatch was established at Northland College in 1978. Innovative for its time, the program used volunteer citizen scientists long before the term "citizen science" was invented.

As the coordinator of LoonWatch, Terry got to know the program's volunteers and had a pretty good working knowledge of lakes that had loons, including nest locations and how many chicks they produced from year to year. The volunteers were called "loon rangers." They were the kind of people whose homes you notice when you drive through lake country because they have loon mailboxes, loon lawn ornaments, and cabin names with loony word play. In other words, the volunteers were loony.

Terry would send them a survey form each spring, and during summer they would collect information on their local lakes' loon populations. In the fall they'd send their data back to Terry and she would tally it up. This data was excellent for tracking loons on lakes where people lived and visited often, but could not be used for a statewide population estimate. To get that number, every five years, LoonWatch conducted a statewide loon survey on a random sample of lakes. This was a one-day survey based on a statistically random computer-generated lake list. On a July morning, hundreds of volunteer citizen scientists would check the selected lakes for loons and chicks, and the data they gathered was used to estimate the number of nesting pairs, chick production, and an overall state population number. Over time, this survey would be used to look for trends in the loon population. Because LoonWatch had such dedicated volunteers, Mike partnered with the program on data collection, citizen science training and public education.

Mike set up his mercury study and hired a field crew. He started collecting samples of feathers and blood and he hired Dave Evers to capture the loons and teach us his new capture technique, the one using chick distress calls to draw in the adults.

By then I had a new boss at Mercer, Greg Kessler. Greg told me if my other duties were caught up, I could work with Mike when needed, especially when he worked in Iron and Ashland Counties. When my work schedule allowed, I started surveying loons. Dave Evers's crew did the banding and processing, and we did the reconnaissance fieldwork to find out which pairs had produced chicks. I took the lead on the TFF. I knew that body of water well, after all. Mike took the lakes in Vilas and Oneida Counties. When Dave's crew, a bunch of 20-year-olds, arrived, they needed housing. My house outside Mercer became loon headquarters for a couple of weeks each year. Tents were pitched in the yard and the beds were full. Sleeping bags were spread on the floors. Boats and trailers filled up the driveway. Mike contracted Dave's crew for three years, then hired his own crew because by then we had loon catching figured out.

The only close call I can remember was when one of Dave's crew, a woman named Jean, dropped a six-gallon tank of gas on a spotlight car battery. Sparks flew and people jumped out of the boat, but luckily the gas tank didn't explode. We had our share of getting lost in the dark, hitting stumps, and breaking down, and once in a while on private lakes someone would see our spotlights and call the game warden to report poachers on the water. We solved that problem by notifying law enforcement about our activities ahead of time. I always went along on the TFF nights in order to lead everyone back to the landing through the dark maze of islands and rock bars at 4 a.m.

Mike researched loons for over 20 years. He's had three graduate students publish master's theses on loons. Among his many accomplishments are the banding and color-marking of over

2,000 loons, contributing to the establishment of new mercury regulations to protect wildlife and humans in Wisconsin, and the creation of an excellent data set on mercury in Wisconsin loons.

Over the years of Mike's research, Terry got more involved in field projects and eventually left her job to become a private consultant working for Mike and others. We had first met during my bear trapping years when I was working with Ray Anderson in the Apostle Islands and Terry was a park ranger on Stockton Island. Working on loons brought us together frequently, and maybe those moonlit nights capturing loons worked a little magic. The loon lady and I became more than research partners—we became life partners. Our individual adventures evolved into adventures together.

A Tenacious Loon

Like people and pets, wild animals have personalities, and loons are no exception. Some loons nest right in front of peoples' docks while others are very wild, fearful, and intolerant of humans. During our years of loon research, we encountered both extremes. We had one loon who nested on a mud flat on the south end of the TFF that was extremely shy. This bird nested in a very remote location behind a number of floating bogs connected by narrow deep channels. When we would visit the nest for monitoring we would never see the birds. We would stop and feel the eggs and note they were warm and being incubated, but we never saw a loon near the nest. He or she would slip off the nest long before we came in view.

Much of the data we have on loons is gathered from time activity budgets (TABs). To compile a TAB, an observer sits for a scheduled period of time and observes and records all the behaviors of the target animal. The specific data Mike Meyer wanted for loons included how much time they spent foraging, feeding chicks, preening, sleeping, incubating, etc. One goal was to determine if or how mercury exposure affected the chicks.

In general, both adult loons will share in incubation, but it's rarely 50/50. Sometimes it's the male that does more of the incubation and sometimes it's the female. They usually switch about every two hours, and we found females doing more nighttime incubating and males more during the day.

One day Terry was doing a TAB session on a pair of loons on the TFF. It was the "Trude West Pair" and they were nesting on a small island across from a campsite. I went along just for the heck of it, and we launched a canoe and paddled over to the campsite where Terry could watch the loon with her spotting scope without flushing the incubating adult. After the two hours of observation was completed, Terry said, "I need to see if it's banded and read the band colors. They haven't nest-switched so I can't tell if it's banded—why don't you paddle over and flush it?" She would zero-in on it with the spotting scope on full power and maybe be able to see the bands when it got up and headed for deeper water.

I paddled over and approached at an angle, keeping her view of the nest clear. The bird didn't flush, so I landed on shore about five feet from the nest. To my surprise, it still didn't flush. I got out of the canoe and slowly crept up to the nest. The bird stayed put, so I started to reach out with my hand, thinking I could lift the bird a bit and see its leg. The loon tried to peck me with its sharp bill. I pulled back and broke off a two-foot-long branch from a shrub that had a forked end. Using the stick to hold the loon's head away, I reached under the bird, feeling its two warm eggs and pulled its leg to examine for bands. I crawled back into the canoe and paddled over to Terry.

"Two eggs, no bands," I said.

"Did you check both legs?"

"No," I said, "just the left."

"Mike puts the USFWS aluminum band on the right leg of adults and the left leg on chicks. You need to go back."

Reluctantly, I repeated the entire procedure. I even had to break another forked stick because I couldn't find the original. The right leg was also unbanded. That was the tamest loon I ever encountered, or more likely, it was not tame, just bonded to protecting that nest. That adult loon could have been predated if it responded that way to a bald eagle, fox, or even raccoon. Mike later told me that out of the thousands of loon nests he had checked, he had experienced several other loons with this tenacious nest behavior.

High-Tech Loons

"We live in a society exquisitely dependent on science and technology, in which hardly anyone knows anything about science and technology."

—Carl Sagan

New Loon Partnerships

In any profession there are personalities and quirky people, but I tend to think wildlifers are quirkier than the average Joe. Terry has a half joking theory that people tend to resemble the animals they study. She points to the wolf and raptor researchers who tend to be more glamorous. Following that theory, folks who study obscure insects or amphibians tend to be quiet and composed. Loon researchers might fall somewhere in the middle, and in my opinion are fun-loving while being serious about their work.

Mike Meyer and Dave Evers got along well and collaborated with lots of other researchers.

One frequent partner was Kevin Kenow, who worked for the U.S. Geological Survey (USGS) out of La Crosse. Kevin's specialty was waterfowl. He had been working on surgical telemetry implants in ducks and had used implants to study migration in canvasbacks in the early 1990s. He and Mike had met as college students. They kept in touch and, building on shared interests in loons and research techniques, began to collaborate on loon work with other researchers in the Midwest and beyond. Mike was interested in identifying causes and timing of mortality in loon chicks. He knew Kevin had developed a subcutaneous (under the skin) transmitter implant technique that worked on canvasback, redhead, and mallard ducklings and asked Kevin to work with him testing the technique on loons. Kevin made some subtle changes to the implant technique and found that adult loons tolerated the same type of egg/chick swap that hen ducks had put up with. The method involved collecting eggs from nests during late incubation and substituting the eggs with water-filled plastic eggs which the birds would continue to incubate.

Like an espionage operation, our study started with surveillance. From the first arrival of the birds in spring, we tracked the nesting pairs, visiting them every couple of days to make sure we found their nests and could identify the first day of egg laying. Loons usually lay one or two eggs, with a day between the first and second egg being laid. Once a pair had nested, we would mark the first egg so later in the season we would be able to tell which was older. Once the pair had laid their second egg, we would slack off our visits, only checking the nest every three days. If a pair only laid one egg, we would drop them from the survey group as we only wanted two-egg clutches for our egg swap experiment. When the pair was a couple of days out from hatch, we would approach the nest and flush the incubating loon, quickly identify the first laid egg, and replace it with a hand painted plastic replica.

We would carefully sandwich the real egg in a cooler with hot water bottles and transport it to our field station where we had incubators set up to warm, roll, and moisten the loon eggs just as if they were in the nest. When the chick hatched, Kevin would immediately implant a tiny subcutaneous transmitter, put the chick in the cooler with its hot water bottle and head back to the lake. We would motor up to the nest, flush the adult and remove the fake egg. Before placing the chick in the nest we would carefully staple a paper towel around it and then gently nestle it into the middle of the nest. Because we always collected the first laid egg, the second would not have yet hatched. The adult would return to the nest and climb up, its wet breast feathers would wet and soften the paper towel and "voila" its chick was "hatched." I am sure that the loons did a real double take when they saw that instant chick.

Kevin and Mike collected some good data on chick mortality as well as testing the research techniques. Two unusual observations were a chick implant signal moving slowly along a lake bottom, probably in the stomach of a snapping turtle or musky, and a second was when a signal was tracked to what was probably the den of a fisher or mink. As part of this effort, Kevin brought six loon chicks back to La Crosse to evaluate transmitter implant effects while rearing the chicks in captivity. Once Kevin was able to successfully raise chicks and Mike saw the excellent rearing facilities available at the USGS center in La Crosse, he approached Kevin to enhance his field mercury study with a laboratory study on dietary methylmercury. Through his years of research Mike's work had been complicated by a number of tricky questions around loons and mercury that were not easily resolved when you had factors thrown into the mix like weather, lakes with differences in food availability and predators, and quirky loon parents.

That first laboratory study at La Crosse led to a larger study where loon chicks were hand-raised there, with feeding and environment closely regulated. Kevin needed to assess survival, mercury levels, immune system responses, and behaviors with just the question of mercury as a variable. We hoped that this additional data could be combined with our existing documentation on wild pairs to fully explore the mercury question. Terry and I assisted with nest monitoring and egg collection on this study.

Over the years we had learned a lot more about mercury and loons. We knew that much of the airborne mercury that arrived in northern Wisconsin came from coal-fired power plants. Sources were both local (Midwest) and global. Downwind lakes and streams are impacted when rain or snow falls, as the precipitation rakes the mercury from the atmosphere. Not all mercury poses an immediate problem, as it is not always soluble in the water column. In many lakes it settles to the bottom and is buried in the sediment. On acidic lakes (with low pH), a bacteria that lives in the sediment "methylates" the mercury, adding a methyl molecule to the mercury. This form of mercury is more readily absorbed by biota and tends to bio-accumulate. Northern Wisconsin has a fair number of these acidic lakes, which are often small and surrounded by sphagnum bogs that can leach acids into the lake system.

These kinds of lakes are also found in the Adirondacks in New York, and parts of New England and Canada.

Mike and Kevin's work compared low pH lakes with neutral pH lakes, as well as looking at loon nest success, chick feeding habits, and other environmental factors that might be impacting loons. They needed the laboratory setting in La Crosse to test mercury exposure in loons without the complications found on natural lakes. With controlled experiments they were able to tease out the story on mercury impacts. It turned out that loon chicks hatched on low pH lakes (where prey fish typically contain elevated levels of methyl mercury) had trouble with thermoregulation (temperature control). They were also slower to respond to threats such as loon warning calls which are typically made when a predator is nearby. Their immune systems were also impaired and there was some evidence that the smaller adult male body size on low pH lakes may have genetic implications for their chicks. Diet was a factor, but the natal lake also made a difference. This was a concern for loon production on those lakes and supported the idea that we should reduce additional airborne deposition of mercury. The good news is we do not have enough of these low pH, high-mercury lakes in Wisconsin for this to pose a threat to the overall state loon population.

Not all the mercury in lakes originates from airborne sources. Some lakes have naturally higher amounts of mercury in the sediments. But the addition of mercury from coal-fired power plants and other human-made sources compounds the problems for fish, wildlife, and potentially humans. One of the difficulties with airborne pollutants is that the toxins don't know where state boundaries lie. This is one reason that state-by-state regulations can only be effective if federal laws set a reasonable baseline standard. These studies were part of the body of science that was used to develop mercury emission policy nationally and in Wisconsin. However, fish advisories still exist on many Wisconsin lakes. Tests on Native Americans in Wisconsin whose diets include higher amounts of fish still show elevated mercury levels.

A New Loon Trap

To stay in touch with the Wisconsin biological community there are a few conferences Terry and I always try to attend. One is the annual meeting of the Wisconsin Society for Ornithology, the other is the Wisconsin Chapter of the Wildlife Society. At the 2007 Wildlife Society meeting, Kevin Kenow approached me saying he had a problem. I love problem-solving, so I was all ears. Kevin said he and Mike wanted to attach a new device called an archival geo-locator tag to the leg bands on loons. This tag was a little over half the size and a third the weight of a triple-A battery. It had enough power to store data for up to two years. The purpose of the geo-tag was to collect date, time, temperature, air pressure and length of daylight at scheduled times throughout the day.

"So, what can you learn from these tags?" I asked Kevin.

He said collecting migration information was his primary interest. He was looking for a less invasive and expensive means of tracking loon migration. It would also give him great data on nesting, foraging, and other behaviors. Since loons only leave the water to nest, breed, and fly, the temperature and date would tell us if the bird was nesting or how often the pair did incubation switches and how long they incubated the eggs. If the bird was on a nest out of water, the air temperature would be different than the water temperature.

Kevin explained that temperature, day length, and date would also give us data on migration. When the loon was flying in the fall the temperature would be air temperature, probably quite different from water. The tags would give a rough idea of latitude from day length and longitude based on the time of solar noon, to within about a 50-mile radius. Stop-over sites, routes, and migration destinations were all likely new information we would acquire along with dates when birds left and when they reached wintering grounds on the Gulf of Mexico or wherever. One last data measurement the geo-tags provided was pressure, which would show how deep the loons' dives were, giving us some clues on feeding habits. If the loon was feeding deep on the seafloor in winter, we could make some guesses about what fish it might be eating.

"So what is the problem you need to solve?" I asked.

"I don't learn a thing," Kevin said, "until I recapture the loon and retrieve the geo-tag to download the information onto a computer." He continued, "After I put a geo-tag on a loon I need to get it back within three years, and one or two years would be ideal. You and I have worked a bit on capturing loons in spring but you will recall how poorly we did and how capture-wise they become. Our current capture technique (the Dave Evers technique) only works reliably when the pair has chicks. Chick production is sporadic due to nest failures from predation or disturbance. Some years 50 percent of the nests fail and these tags cost $1,000 each. It's important we figure out how to retrieve the tags the following year even if the loons don't have chicks."

"I'd also like to catch some adult loons on territory in the spring before they nest," he said. "I want to blood sample pre-nesting loons so I can determine their mercury levels just after they arrive back from their wintering grounds on the ocean. That way if they have chicks, we could recapture them and measure the mercury they acquired from the time they arrived on Wisconsin lakes. This would help us understand what proportion of the mercury burden might be coming from the wintering grounds versus our nesting lakes."

I scratched my head on this one for a while. I'd first gone through all this back in the 1980s, working with Jerry Belant trying to devise loon capture techniques.

Kevin and I started brainstorming. We had been experimenting on ways to capture loons without chicks and prior to nesting for a couple of years with minimal success. Kevin had modified a loon decoy, placing a small tape recorder playing territorial calls on its back and using the decoy as a lure we could anchor in the middle of a floating mist net. The net was held afloat by a ten-foot diameter plastic hula hoop-like tube. We had set this prototype afloat in deep water within loon territory and watched what happened from a distance. We attracted some attacks to the decoy and we did catch a loon or two, but this technique had major drawbacks. First, it was very difficult to transport in a boat and deploy due to its large size. In addition, the net would get caught on anything and everything in the boat and we were constantly untangling it, making setup a nightmare. Once deployed, our challenges continued. If a loon attacked, the decoy would flip, ruining the tape recorder which cost $30 to replace. We got way more loon refusals than attacks, and I think this was because the black hula hoop pipe was too visible. We eventually abandoned this technique.

Suddenly, I realized I had been wool gathering and Kevin was looking at me expectantly. We decided to adjourn to a quiet corner in the conference hall and started brainstorming in earnest. We recalled previous failed attempts from Jerry Belant's research and my modified Bailey beaver trap device to our own efforts from previous years. Kevin and I pondered the problem, turning it over and looking for a new twist. Then a lightbulb went on; what about an underwater-lift net? Nets for capturing birds were not new to us. When I worked in Park Falls in the 1970s, I'd caught dozens of woodcock in Japanese mist nets, which look like badminton nets with cobweb thin mesh, used to safely capture birds for banding. Terry and I had mist-netted eider ducks in Manitoba, several species of river birds in Nepal, and golden-winged warblers in Honduras. Kevin had decades of experience capturing waterfowl with dip nets, bait nets, and decoy traps.

I'd also observed how the DNR fish crew caught sturgeon in underwater gill nets. The nets were suspended in the water from floats and when a sturgeon swam into the net it only passed partially through before becoming entangled by its gills and fins. The fish crew watched the nets and when they saw the floats bobbing up and down they quickly pulled the nets up into the boat and untangled the sturgeon for tagging and release.

We knew loons were very territorial and defensive when they arrived at their breeding territories in spring, and an intruder loon decoy in their territory would really rile them up. We decided to try an anchored loon decoy out from the shore in about three or four feet of water to get the loon's attention. We would string out 100 feet of sturgeon gill net in a horseshoe shape from shore in front of the decoy, anchoring the net by driving four pipes in the lakebed. Our plan was to attach eye bolts and cord on top of the pipes for lifting the net. To set the trap we would lower the net under water and push it to the bottom so it was inconspicuous to the loons and then hide on shore and play territorial defense calls, like Jerry Belant had done to entice the loons to attack his decoy. When

the pair moved in to attack with the target loon clearly inside the net, Kevin would pull the cords and lift the net. The startled loon would dive to escape and become entangled in the net. We would then run out in our chest waders, untangle, and retrieve the captured loon for banding.

"I think we should try it," Kevin said.

I told him there was plenty of used pipe at the Mercer DNR station and Bruce Bacon had a post pounder we could use to drive in the lift pipes. Jim Cox, the fish technician, had a net catalog, and a spool of cord for lifting. When we later consulted with him, he suggested we order the size they use for gill-netting musky. I told Kevin the ice should be out on the Flambeau in about the middle of April and the loons should start showing up on their territories shortly after that.

Kevin arrived in Mercer a few days after ice-out and we started building the net trap. We headed to the lake and tried the lift net trap. Boy, did we catch loons! In fact, we got so many doubles (both birds in the net at the same time) that we added a third member to the catching crew to help untangle loons. On one double, the water level was close to the top of my waders, and when I lifted the net, the loon was level with my face. It beaked me between my nose and my eye. A loon's bill is like an ice pick, and that stab wound was a close call. By the end of spring banding season our hands were a patched-up mess of wounds. We found that handling loons in the daylight was a lot harder than at night. The loons could see and aim better with their sharp bills.

We did some tweaking of the technique. Kevin made improvements on the gill net mesh size for loons and added a lead line to the bottom of the net. We also missed a few birds because on clear lakes they could see the net on the lake bottom and refused to swim over it into the trap. We solved this problem by dyeing the monofilament net brown with fur trapper's dye. Kevin wanted to publish this technique, so he wrote it up, putting me down as co-author, and Terry did the illustrations. We published our results in the Journal of Field Ornithology in 2009.

A few weeks after our article came out, Kevin got a call from Dave Evers, who now worked through the Biodiversity Institute in Maine. Dave told Kevin their crew had tried our lift net trap (we had shared our design with USGS staff in Anchorage, Alaska, in 2008 and word travelled about the technique) and it had worked great. He mentioned he had thought of the idea too, but we beat him to development and publication of the technique. I told Kevin, that's interesting, Jerry Belant and I had just missed figuring out Dave's capture technique (see "Close, but no Cigar"), and now the shoe was on the other foot. Kevin even went a step further with making this loon-catching technique high-tech. He disassembled a remote-control goose hunting decoy and put the parts inside a loon decoy. This battery-powered loon with an electric propeller could be sent out to circle around the live loons to entice them to follow it back it into the capture net. Kevin called it Trudy. We had quite the setup. I have since heard back that this technique is being used in Alaska where night-lighting doesn't work well because they have 24-hour daylight much of the summer. One researcher there caught over 20 loons using our technique.

A Covert Operation

In the winter of 2009, Kevin called to ask if Terry and I wanted to work for him on a contract again that summer. He and Mike were going to use satellite telemetry devices that would be surgically implanted inside a sample of loons. The technique had been successfully tested by Kevin, and the battery transmitter would be implanted inside the body cavity with the antenna emerging low on the loon's back just above the tail. The transmitter sent a signal to a satellite so he would be able to track birds on his computer all year long.

I was intrigued. Back in the 1980s, Jerry Belant had tried some radio telemetry devices on loons without success. We'd discovered that due to the lifestyle and physiology of loons, they're a difficult species for attaching external telemetry transmitters.

Consider these facts. Loons can dive deep (over 200 feet) and can fly 70 mph, but to get airborne they need a quarter-mile runway. Unlike most birds, loons have solid bones rather than hollow ones. Most birds are adapted to be light for flight, but for loons there must have been an evolutionary tradeoff between flying and diving. Solid bones aren't as susceptible to being crushed by underwater pressure. Being heavy helps loons dive deeper, but that also makes getting up in the air a challenge. Unlike many birds, loons can't soar or hover and have to flap continuously to stay airborne, so adding more weight with a telemetry device could make an already difficult take-off even harder. The transmitter's weight was well below the "five percent rule" recommended by the Bird Banding Lab.

Loons also have thick, dense, oily feathers to keep them warm and dry while on the water. They preen frequently, adding oil from a gland located near their tail to keep them water-sealed, so any kind of harness you put over the plumage could cause a leak in the feathers and get the loon wet and cold. Jerry Belant tested a few different harnesses and a poncho contraption for carrying telemetry transmitters, but in each instance the loons got wet and had trouble thermo-regulating so the equipment had to be removed. The backpack harness style has been used on hawks, owls, woodcocks, and ruffed grouse, but was no good for loons. Plastic neck collars like those we put on geese and swans won't work either, because loons need to be able to expand their necks to swallow large fish. The collar might cause them to choke, or if it was big enough to allow for swallowing, it would probably fall off.

During Mike's loon mercury research project, questions had arisen such as potential for sources of mercury during migration. Kevin said that his technique would allow them to track loons during migration and wintering. To date, records on migration routes were mostly from recovery of banded birds that died. This technique would help us more clearly define migration routes, wintering grounds, potential threats, and critical habitats.

"This internal implant sounds risky," I said. "How do you safely implant the transmitter?"

Kevin explained that he had first developed the surgery techniques for implants on waterfowl and then tested it on loons, conducting the surgeries himself. Later, he had trained a veterinary surgeon with USGS in Alaska who shared the knowledge with other veterinary surgeons and they had lots of experience. One was a vet from the University of Florida, and Kevin was bringing him in as part of the team.

Terry and I were both a little queasy about this method. On a gut level it felt a little scary. Over the years we had developed quite an affection for our research birds. We always felt a little guilty about all the interruptions we caused our loons as we counted eggs, conducted time activity budgets, and captured and banded the birds. We always did our best to minimize our impacts, but in reality, any time you interfere with a wild animal you have to accept the fact you could inadvertently harm it. For that reason, methods were carefully scrutinized and protocols were set up to reduce impacts as much as possible.

When we brought this up, Kevin explained the rationale for this project and his protocols in more detail. His testing included a three-year study in New York, Maine, and New Hampshire that developed protocols. He told us that the previous fall, during migration, hundreds of dead loons had washed up on the Lake Michigan shore, especially on the northern shores including Sleeping Bear Dunes National Lakeshore. The necropsies of the dead loons showed they'd died from Botulism E. Their digestive tracts contained an invasive fish called the round goby, and it was suspected that gobies were getting the botulism from eating zebra mussels, another invasive species common in the lower Great Lakes. These two invaders had entered Lake Michigan in the ballast water of ocean-going cargo vessels. Kevin and Mike wanted to collect more data on the distribution and foraging patterns of loons and other water birds related to these botulism die-offs. They felt that the satellite data would help to pin down these tricky migration questions.

Terry and I offered to help, and the first step was to locate birds for the study. Kevin had 10 radio transmitters to implant for the pilot project (in Wisconsin and Minnesota). He thought that the TFF would be a good research site because we had as many as 32 pairs of loons and could easily find seven pairs with chicks, which would mean we could be sure to catch the adults. He only wanted to put the transmitters in males because he was not sure whether the abdominal implant would affect egg development in females, and he didn't want to risk losing fertility in a bird. We would still put a geo-tag on the leg band of the females. He also said that the TFF would be a better place to do this work because we would be less in the public eye if something went wrong and a bird was injured or, worse yet, died. As much as he hoped this was a remote possibility, he didn't want to impact a lake with only one pair of loons.

That spring, Terry and I started checking our loon territories, and by June had found seven pairs with chicks. We informed Bruce Bacon, as well as the local DNR warden. Mike was monitoring pairs, and the biologists in Minnesota were also tracking candidate loons. The first surgeries were done on Minnesota birds. For the Wisconsin implant candidates, our crew did the captures with Kevin's USGS staff and the veterinarian managed the surgeries. Kevin wanted Terry to drive the capture boat and he wanted me on shore to help with handling. Then he asked if I could find a place near the loon pairs to park the FEMA trailer. "FEMA trailer?" I asked, "Aren't they the small trailers the federal government gave people to live in down in New Orleans after Hurricane Katrina?"

"That's right," Kevin said. He explained that USGS had one converted into a medical surgery lab. The trailer was 30 feet long, set up with bright lights and running water, and it had a gas generator that ran the power. Could we find a parking spot where they could park this trailer, as well as their other trucks and boats? They thought they'd start at 10 p.m. and should be done by five in the morning.

I thought a minute. Mercer is a small town of 800 people, and people love to gossip. People also love conspiracy theories and I started imagining what some of the locals would think if they drove by our operation in the middle of the night, seeing not only a big trailer with U.S. GOVERNMENT, FEMA written across it, but a bunch of trucks with federal government license plates, and people with headlamps. Even worse, what if they looked in the trailer window and saw people with surgical gloves and masks doing surgery. I had a vision of another Roswell, when an urban legend got going that the government was covering up a Martian landing or UFOs.

My fears were groundless, and all went well. It took two nights to implant the transmitters in seven loons, and Terry and I monitored them during their recovery. The seven males were back with the females and chicks on their territory the next day and they all survived. The following fall and spring, Kevin followed the loons on their migration, even setting up a website for citizens to follow the loons online. Most of the loons used Lake Michigan as a stopover site, as Mike and Kevin had suspected. It was fascinating to watch via computer as they hopscotched their way south. Most went to the Gulf of Mexico, a few to the Atlantic coast, and one even wintered in open water on a lake down in Indiana.

The next spring Kevin and I watched as the birds migrated back and we re-caught them with our lift net trap. We removed the geo-tags and examined the birds. I was amazed at their health and vitality. The antenna exit location was completely healed and fused to the body as if the antenna had been there all along. That summer we implanted transmitters in an additional five birds in Wisconsin, 13 in Minnesota, and three in Michigan. Each year's data was made available online so the public could click on each bird and watch its travel route. It was especially interesting to see how quickly the loons migrated from southern Lake Michigan to the Gulf Coast, sometimes traveling the distance in just two or three days. At one point, the project became really popular with the locals, and some of the taverns in Mercer ran a lottery to see who could guess which loons would return first.

The batteries continued to transmit for up to 24 months, and at this writing have long since gone silent. Last spring, Terry and I saw three of the "antenna birds" still in their old territories. The antenna is still visible with binoculars. Nobody in Mercer ever noticed our two-night covert operation (https://umesc.usgs.gov/terrestrial/migratory_birds/loons/migrations.html).

Kevin's loon satellite telemetry data ended up serving Wisconsin, Minnesota, and Michigan in an unexpected way. On April 20, 2010, an oil rig, now known as the Deepwater Horizon, blew up, spilling 4.9 million barrels of oil into the Gulf of Mexico. Dave Evers's crew in Maine (Biodiversity) was awarded the contract from the federal government and British Petroleum (BP) to study the impact of the oil spill on loons. Because Kevin had expanded his migration study into Minnesota and Michigan, his satellite telemetry research was used to document the extent to which Great Lakes loons wintered in the area of the oil spill. Minnesota was in line to receive economic compensation from BP for Midwest loon losses, with the hope of further protecting loon nesting habitat and replacing the birds lost to the oil spill. I suspect Wisconsin missed out on that pot of money because the political winds had shifted, and research was no longer a priority for our state administration. In 2015, the Republican-controlled government reorganized the DNR and gutted the budgets, eliminating most research scientists and focusing wildlife work on game species. Without a reliable state agency partner, the federal government opted to send those BP funds to a state more prepared to value its loons.

Pink Rock Rescue

It was the middle of April and my wife Terry was visiting friends in upper Michigan for the week. Kevin Kenow and his assistant Steve Houdek had come up from La Crosse to do some early spring loon catching and retrieving of geo-tags. We were working on the TFF, which is a fairly remote, wild place any time of the year, but in April it is virtually deserted. Fishing season doesn't open until the first weekend in May, and April is too cold for camping. When we left the dock that spring morning for a day of loon capture, we took lunches and extra clothing. I'd left my cell phone behind. I tend to forget my phone and, besides, I thought it might get wet. April can bring some tough weather and we can have snow as late as May. The day was brisk and the water was mighty cold—ice had gone out only the previous week.

The day went well, and about an hour before dusk we decided to try to catch one more bird, the "Pink Rock" male. We often named our pairs after a feature of the lake in their nesting ter-

ritory. In this case there was a big pink rock—rose quartz—that projected out of the water like a small island about the size of a Greyhound bus. The previous year a pair of loons had nested off a little gravel shoal at the base of the rock and produced chicks. The Pink Rock male was one of Kevin's satellite telemetry implant birds, and we needed to catch it and remove the geo-tag attached to its leg band.

When we got there to set up our capture nets, we picked a location on a small, one-acre island 100 yards from Pink Rock. We landed and set up the lift net trap, and Kevin and Steve hid in the shrubbery. After everything was set up, I couldn't find a good place to hide the boat, so I left Kevin and Steve to do the catching, and I motored over to Pink Rock and hid myself and the boat behind the rock. Watching through binoculars, I could see perfectly.

They started blasting out loon calls. The loon saw the decoy and responded, heading for the decoy and net. As I watched, the net came up and I could see splashing. Kevin and Steve ran into the water to retrieve the bird. I jumped in the boat to head over and help. Now, I didn't tear out of there at high speed—I backed out, turned the boat around and steadily increased the throttle. The problem was that it was almost dark, and I couldn't see into the water very well, so I ended up taking a slightly different route than when I'd pulled in. A moment later...WHAM! The boat lifted a little and stopped. I had found Pink Rock's sister. This rock was four inches under the water and about the size of a Volkswagen beetle.

I wasn't worried. The TFF is full of stumps and rock bars, and I had been stuck on plenty of them. Usually, all you have to do to free yourself is move all the way to the other end of the boat, put the motor in reverse, and give her the throttle. The combination of weight shift and thrust allows you to back off the underwater obstruction.

This time that technique didn't work. I was high-centered on that rock with the aluminum hull of my boat. It bobbed like a teeter-totter. I tried everything. I stuck an oar down but couldn't touch bottom. The water, it turned out, was over 10 feet deep everywhere around the boat. I had on chest waders, so I tried getting out on the rock to push, but it was slimy, slippery, and slanting. I could not get a firm boothold to push from.

By now Kevin and Steve were done processing the loon. They stood on the shoreline of the island watching me rev the outboard and spinning around in circles trying to get off the rock. It was getting darker, and I realized I had their extra clothing in the boat. As I tried different escape tactics, I thought about the fact that nobody knew where we were and it could be days before anybody casually came by, perhaps not for two weeks until fishing season opened. I gave up my attempts to escape. The boat wasn't moving and it was now dark. I yelled at the top of my lungs to Kevin, "Do you have a cell phone?"

"Yes," he yelled back. Kevin's island was about 100 yards away and I could hardly hear him.

"Do you have reception?" I called out.

"Yes," he said, "one bar." This meant his phone would barely reach the cell tower to transmit a call.

"Call my son, Zach," I yelled. We could barely hear each other due to the distance, and it took about four tries before he got the 10-digit number right.

"Where do I tell him we are?" Kevin yelled.

"Just tell him Pink Rock. He'll know."

Luckily, the call went through. About two hours later I saw a faint light on the horizon. It was pitch dark by then, but I could see that the light was getting closer. The TFF had hired a new seasonal ranger for the summer named Kenny Pemble. Kenny had just that day gotten his ranger boat out and prepared it for the upcoming season. Kenny was a friend of Zach's and grew up in the house next door to ours. Boy, were we glad to see them. They got to me first, hooked a rope to the boat, and pulled with their 90-horsepower outboard. It took them three tries to get me off that rock, then we retrieved Kevin and Steve from the island. On the way back to the dock Kevin said he had considered leaving his phone in the truck, not wanting it to get wet.

Innocent Until Proven Guilty

Wisconsin's common loon population resides in the northern third of the state, except for a small pocket in the Central Sands area where there are a number of commercial cranberry operations with related lakes and wetlands. Michigan and Minnesota are similar in that loons nest in the northern parts of both states. Historical records indicate, however, that common loons once nested as far south as Iowa, Illinois, and Indiana.

In our study area in northern Wisconsin, nest success varied considerably from year to year depending on a number of factors. Some years 50 to 80 percent of the nests failed due to black fly population explosions. You might be familiar with black flies as pesky biting insects that torment you during May and June fishing and camping trips. Loons suffer a similar plight as there is one species of black fly whose life cycle is tied to loons. This black fly attacks loons, seeking a blood meal at the time when loons are incubating on the nest. The fly only feeds on loons and is attracted to the oil the birds use to preen their feathers. The flies tend to be worse on lakes with boggy shores, and sometimes they can even be seen hovering over old nest sites, apparently sensing the oil gland excretion even though the nest is empty. The flies torment nesting loons, and sometimes the birds' heads are completely covered with the biting insects. Loons respond by leaving the nests for relief and sometimes even abandoning their eggs. Loons aren't alone in their suffering—a similar black fly is the cause of much of the whoop-

ing crane nest failure at the International Crane Foundation's recovery project near Babcock, Wisconsin.

Fluctuating water levels during high precipitation years can also affect nest success. Loon nests are often built just a few inches above the lake level. This helps the loons move onto the nest, as their legs are located far back on their bodies and they have to push themselves on their bellies onto land. Nesting close to shore near the water level makes for easy loon access, but also makes the nests vulnerable to rising water from heavy rain, flooding caused by water control structures, even high-speed boat wakes.

Predation is also a major factor affecting nest success—we often found predated nests. Sometimes there were clues left behind such as telltale eggshells, but often the eggs were simply gone. We'd do detective work, a.k.a. "Loon CSI," by looking for tracks or scat or cached eggs in the area to determine what had happened.

We wondered how well loons could defend nests, using their sharp bills or attempting to scare away a threatening predator. We'd have these discussions after work over a beer, or while waiting at night for the sky to get dark enough for capturing. When a pair lost its nest or had a particularly odd predation, we would throw ideas back and forth. Otter always came up, along with raccoon, mink, fox, gull, crow, and bald eagle as possible suspects. River otter are abundant throughout northern Wisconsin. We'd often see otter while watching loons, and their scat was often found at communal otter "toilets" near loon nests. Way back in 1952, Sigurd T. Olson, the son of writer and wilderness advocate Sigurd F. Olson, did his master's project studying common loons in the Boundary Waters Canoe Area Wilderness in Minnesota. His thesis suggested river otter could be predators of loon nests. From that point on, many loon researchers referred to his statement as evidence, but I could never find another study that had conclusive evidence proving that otter were loon egg eaters. I thought they were innocent, but I was in the minority.

A complicating factor for nest success was that the landscapes in the Midwest had changed a lot over the past 100 years. At the current time, Wisconsin is the southern edge of loon nesting range in North America. It is a generally held belief that species are most vulnerable along the edge of their range. If loon range used to extend much farther south, what was limiting them from spreading further south now?

When you examine habitat availability for loons, it is clear that shoreline development has skyrocketed in the last 50 years, putting stressors on available nest sites. With development came more boating, fishing, and waterskiing, and in the 1990s, a new invention called the jet ski started creating tremendous waves. Had these changes affected loon habit enough to shrink their historic range?

Lakeshore development also brought gardens, garbage cans, fish cleaning, birdfeeders, and woodsheds. All of these provided perfect habitat and food for a modest-sized egg-eating mammal, the raccoon. If a natural undeveloped lake could support a certain number of raccoons, could a developed lake support more? If raccoons increased, how was loon nesting success effected? Historical records showed loons hadn't evolved with raccoons on the scene. In pre-European settlement times, raccoons had only lived in the southern states and Mexico. In the 1940s, the Wisconsin Conservation Department introduced raccoons as a furbearing species. For several years after the introduction, raccoons were protected from harvest until their populations were well-established.

Other landscape changes in southern Wisconsin, Illinois, and Iowa also had potentially negative effects on loon habitat. Many of the prairies and savannah forests were converted into farmland and urban areas. Iowa has a huge raccoon population for one reason—corn. Along with tilling up historic prairies, Midwest farm states also drained many historic wetlands. Both of these practices contribute to significant issues with agricultural runoff and sedimentation of lakes and streams. This creates lakes with murky water not well-suited to loons, who are sight-feeders and require clear water to see fish.

In 2004, Mike Meyer decided to investigate one of these issues: changes in loon nest predation. He had two of his assistants, Nick McCann and Dan Haskell, use some new technology to do surveillance on loon nests. They set seven trail cameras on nests to monitor predation, but it was challenging to find a location where the camera would record a predator without recording every movement the loons made. Nick and Dan decided to focus the camera behind the nest site, which worked well for terrestrial predators, but they missed avian predators or anything approaching from the water side of the nest. Regardless of these problems, this pilot study confirmed that raccoons were a common predator of loon nests.

In 2012, when Bruce Bacon and I were both retired, we decided to apply for a small grant to expand on Mike's predation study. Mike gave his blessing and we applied to the LoonWatch program and to the Wisconsin Society of Ornithology for grants. We received both grants, bought eight cameras, and borrowed two more. Camera technology had improved a lot since Nick and Dan's project eight years earlier, and our new cameras could be pointed directly at the nests. Their memory cards could hold thousands of pictures and their battery life would last the entire 28-day incubation period.

We rotated the cameras between 21 nests on the TFF and hired a student intern from Northland College, Steve Thielking, to help with the fieldwork. He lived at the cabin with Terry and I, moving into our guest cabin for the six-week study, and later compiling the data for publication. As it turned out, we picked an unusual year. It rained a lot, causing the water on the flowage to rise eight inches during nesting. We lost many nests to washouts and wave action, but we improved our camera placement methods and even photographed loon eggs being washed away while a frantic loon tried to build up the nest against the flood.

This data was helpful in documenting impacts of water level management at the Flambeau dam, which controls outflows on the TFF. We also documented some predations, mostly raccoon and even a mink. One day Steve called me, all excited. While he was scrolling through a camera card, he had found a picture of an otter standing over a loon nest with two eggs in it.

The series of photos showed the otter flushing the nesting loon and then crawling up right next to the eggs. To our surprise, the otter left the eggs and never returned, and many days later the nest hatched. While this is only one example, and one example doesn't make a scientific study, I felt vindicated. That otter could have returned and gotten a meal at any time if it wanted to eat loon eggs, but it didn't. To this day we continue to debate the otter question amongst the loon crew, but I still think otter are innocent.

Our study objectives were to document common nest predators, but we also added supporting data to the established argument that water level fluctuations during nesting can have significant impacts on nest success and that those who regulate dams on loon nesting lakes should consider moderating water level fluctuation during the nesting season.

Economic Decisions, Claire, and the Butterfly Effect

"To be truly challenging, a voyage, like a life, must rest on a firm foundation of financial unrest. Otherwise, you are doomed to a routine traverse, the kind known to yachtsmen who play with their boats at sea—'cruising' it is called. Voyaging belongs to seamen, and to the wanderers of the world who cannot, or will not, fit in. If you are contemplating a voyage and you have the means, abandon the venture until your fortunes change. Only then will you know what the sea is all about."

—Sterling Hayden

The Gap Year

Unlike the old French voyageur quoted at the beginning of this book, I didn't spend all my money on pleasure, but being a voyageur at heart, I made some decisions favoring adventure over financial gain. Three of those decisions were ones my parents often pointed to when offering me unsolicited advice.

For instance, building a cabin on an island in the middle of a lake three miles from the nearest landing might not have been a move approved by a financial planner, which, luckily, I didn't have. Clearly, it would have been cheaper (and a lot less work) to have bought land on the mainland and built a cabin with a road, electricity, and vehicle access. As I write and gaze out my window, however, I can see three otter, sliding and cavorting across the ice. Out this same window I've watched a pack of six wolves tracking prey, mink feeding on crayfish, loons incubating a nest, and sandhill crane chicks swimming across the bay. Yes, that's right—sandhill crane chicks swimming. That was a comical sight as, unlike waterfowl, they aren't equipped for swimming. These are just a few of the incalculable benefits of living in a remote, wild place. I also would not have become friends with half the crazy people I know if I hadn't built on this island, including my wife.

Another questionable move I made was retiring early, at age 50. To do that, I transferred from my job as a wildlife technician and became a forestry technician. The physical demands and dangers of jobs like conservation warden and forestry/firefighter earns "protective status," and allows for an earlier retirement. This ensures that the state doesn't end up with a bunch of old-timers running around fighting fires and chasing bad guys with guns. Firefighting was challenging work. I ran a bulldozer pulling a fire plow, and my job was to plow a furrow alongside or in front of the fire, removing burnable material and creating a fire break so the fire would burn itself out. Sometimes I'd have flames right beside the dozer. I was only on fire control for two seasons but the last summer included fighting fire in Montana, which was quite the adventure.

Because I retired early, my monthly annuity was greatly reduced from what it would have been if I'd worked until age 65. But how much money does a person need anyway? After retirement, I took up part-time contract work to supplement my income, which provided us with flexibility to go adventuring during our time off.

Perhaps my most economically questionable but experientially-rich decision was what we called our "Trip-Around-the-World Year," from June 1995–1996. That year my son Zach turned 18 and graduated from high school, and we decided he needed a "gap year," a year off between high school and college to travel, see the world, and, ideally, acquire inspiration and wisdom. Terry, my girlfriend at the time, and I decided we needed a gap year, too.

When I was 19, I had joined my cousin Suzy on a summer trip around Europe. She'd been over there before, but the parents all thought she shouldn't go alone, so they convinced me to accompany her. We traveled through Spain, France, England, and Norway.

I went reluctantly, not wanting to leave my friends and outdoor pursuits, but by the time I got home I was a changed man. The whole world had opened up to me.

Thinking about that formative experience, I dreamed up the trip for Zach and started lobbying. I argued that Zach had lived in Mercer with me since he was in kindergarten, spending summers and most school holidays in Iowa with his mother. We'd gone on a couple of canoe trips in Minnesota, and we had hunted, fished, and trapped together, but we hadn't ever taken a real vacation. Now he was planning to go off to college, and I didn't know where he'd end up after that. I wanted to give him a traveling experience like the one I'd had in my youth. Secretly, I was also ready for a break from my work routine.

For Terry's part, she had worked a variety of seasonal positions for the Apostle Islands National Lakeshore for eight years, followed by grad school, and eight more years as staff biologist for Sigurd Olson Environmental Institute. She was ready for a change of pace as well.

We'd been talking about the three of us taking this trip when I heard about inexpensive "around the world" airline tickets on public radio for as low as $2,500 each. According to the travel agency expert, you had to go in one direction around the planet, and the number of stops was limited. We started talking about how we could pull this off and decided that each of us would save up the money for the plane ticket. Zach had worked construction jobs each summer in Iowa and had saved much of the money already. Terry and I started saving, and we soon had the plane fares covered.

My plan to cover the bigger living expenses was to mortgage the Mercer house for $20,000 and rent it out during our year abroad. The rent payments would go to the bank to pay my mortgage. If we divided the money equally, we would have $6,666 each to live on. This would average out to only $20 each per day, but we thought if we picked our destinations carefully, camped whenever possible, and budgeted, it might work.

The next problem for Terry and me was getting a year off from our jobs with a guarantee we could get them back when we returned. Luckily, the ranger station had an excellent seasonal employee named Steve Kangas, who had been working for us every summer while in college. He knew my job inside and out. With some shifting around of workloads, my boss, Bruce Bacon, and his boss, Fred Strand, found a way to grant me a leave of absence, supporting our plan to do international wildlife projects and broaden my education.

Terry's supervisors at Northland College also said she could have a year off without pay and get her job back when we returned. We started investigating projects we might work on as we traveled, volunteer organizations we could join, and the logistics of getting mail and funds transferred while we were abroad. Cell phones and email were not widely used in 1995, and most places we visited didn't even have ATM machines, so we arranged to visit American Express offices in each country to pick up mail and get cash about once a month. We also had to figure out what equipment and clothing to take for a year in varied climates, from campsites to urban settings, and fit it all into backpacks we could carry.

Our decision to make this trip around the world didn't just affect our three lives and those who knew us. I like to think of it as an example of the butterfly effect, a term coined by Edward Lorenz, a mathematician and a prominent creator of the chaos theory. He wrote, "Does the flap of a butterfly's wings in Brazil set off a tornado in Texas?"—suggesting the actions of one thing can have unexpected effects on another, perhaps half a world away. His theory suggests that everything is connected, and if you look closely, you will find relationships and linkages everywhere.

Our butterfly effect began shortly after we set off on our trip around the world, when a young woman named Claire Gower, from Shalsord, England wrote a letter to the Mercer DNR Ranger Station. It was addressed to wildlife manager, John Olson, but because John had transferred to La Crosse five years earlier, Bruce Bacon opened it.

Claire's letter stated that she was majoring in Wildlife and Countryside Management at a small University of London-affiliated college and would like to volunteer to work a year abroad studying North American wildlife. Her professor had shown her a copy of an old 1984 Wisconsin Natural Resources Magazine that had an article in it written by then wildlife manager, John Olson. It was about co-management of wildlife on the TFF between DNR and Northern States Power, the owners of the flowage at that time. It showed pictures of osprey, eagles, and loons, and views of the vast lake area with its 195 islands. Claire thought, "Wouldn't it be wonderful to work there?" In 1995, correspondence between the U.S. and England required either a long turnover in letter-writing or an expensive phone call. Our state had no protocol for accepting volunteers from foreign countries, but Bruce thought what the heck, he could use some help. Bruce knew this young woman was inexperienced and there was no way to even interview her. He also knew that if she arrived and things didn't work out it might be difficult to get rid of her; after all she was a volunteer. Then there were the questions of where she would stay and how she would be able to afford food. He thought he would need to find at least a stipend to pay her for living expenses.

Bruce took the chance, and Claire was invited to come to Mercer for a year. A month after I left, Bruce went to the airport and picked up Claire, a 21-year-old freckled, red-haired, charismatic woman with a pronounced British accent. The first flap of the butterfly effect had been put in motion. There would be more to come later.

Now I can't possibly tell you everything that happened in the year we spent on this world trip. That would take another entire book. We tramped through Canada, Alaska, New Zealand, Australia, Bali, Lombok, Thailand, Nepal, France, England, Italy, Guatemala, Honduras, Belize, and Mexico before dragging ourselves home.

To cut expenses we volunteered on various wildlife projects, which usually provided housing and sometimes food. We joined an organization called Willing Workers on Organic Farms (WWOOF), where we worked on farms part-time in exchange for room and board. We delivered a van across Australia for a rental company, rode on horses, elephants, bamboo rafts, bicycles, helicopters, tuk-tuks, rickshaws, bemos, a tundra buggy, dugout canoes, ships, boats, buses, and airplanes, and walked through tundra, jungle, mountains, and down riverbeds. We also stayed with friends and relatives and relied on the kindness of strangers.

Our wildlife projects involved bush-tailed possum in New Zealand, loons in Alaska, river birds in Nepal, and coatimundis in Guatemala. But first, starting close to home, were snow geese in Canada.

Transitions in the Tundra

We put the word out to our friends and colleagues that we were looking for projects to dovetail with our travel plans, and the first to come through was Dr. Don Rusch, a professor at UW–Madison. I had worked for Don on a ruffed grouse project in Chatfield, Manitoba and I had contacted him to see if he knew of anything we could do in the Arctic. At the beginning of Don's career, he had been a biologist for the province of Manitoba, and currently had a long-term Canada goose project based out of Churchill, a small town up on Hudson Bay. Don's research station was called "Nester One," and he didn't have any room for volunteers that season, but about 20 miles away was another research station called "Nester Two," run by Don's good friend and colleague, Dr. Robert Rockwell (Rocky) from New York University and the American Museum of Natural History. Don put us in contact with Rocky, who invited us to join his crew on the tundra for a month studying snow geese. He would provide room and board and transportation from Churchill to Nestor Two. In early June, we packed up and headed for Canada, the first leg of our world trip.

We first realized we were really at the ends of the earth when we took the two-day train ride from Thompson, Manitoba to Churchill. We had already driven two days north from Wisconsin, and for hundreds of miles all we saw were spruce trees, creeks, undeveloped lakes, and bogs. Then we hit Thompson, a kind of jumping off place for the hinterlands where the road ends. There we boarded a train often referred to as the Polar Bear Express. The train ride was slow, as the engine and cars wobbled along tracks laid over uneven permafrost. The trip took us through the transition zone from the boreal forest into the treeless tundra. As it traversed the landscape it would periodically stop and a few, mostly indigenous, people would leave the train and walk into what looked like wilderness with only a rough trail or ATV track. The hours passed, and finally we started to see ramshackle buildings, and Churchill appeared. After stepping off the train, we met Rocky and some other biologists who had arrived the day before, and started the next leg of our journey. We drove to the Churchill Northern Studies Center, a research facility 10 or so miles from town, and then began the 20-mile overland trip to Nester Two by tundra buggy. Now, a tundra buggy is quite the machine. It has six-foot-tall tires and a cab and truck bed on the back for storing gear. We rode in the cab and bounced along for hours as the driver skillfully maneuvered around boulders, through streams and up glacial eskers. The flat tundra spread around us, and the sky seemed infinite. Birds flew up from the short Arctic shrubs and water filled potholes. After several hours we arrived at Nestor Two a bit battered from the bumpy ride and ready to stretch our legs. A month later, on the way back, I actually drove the buggy. It reminded me of my tractor days on the Iowa farm, but on steroids.

Nester Two was a series of Quonset huts, connected by rough planking set on the spongy permafrost—a cookhouse, bunkhouse, outhouse, and storage buildings. Located just inland from the shore of La Perouse Bay, part of the huge Hudson Bay lowland, the camp was set along the Mast River Delta, a series of braided streams and willow tundra that covered thousands of square miles. It was early June, and everywhere you looked you could see water, shrubby willows, and permafrost vegetation peeking out from the remaining snowbanks. That first day it looked cold, damp, and bleak, and the rough buildings seemed muddy and slapdash. Terry was particularly skeptical, as there were only a couple of other females at the camp and a fair amount of ribbing and rough language amongst the wildlife "guys." The camp was shared between Rocky's wildlife researchers and a botany crew from Canada, who seemed a bit less rough around the edges. Once we settled, camp became home, and the comforts of hot coffee, cook stove and a homebrewed beer at the end of the day were perfect rewards for the hard work, not to mention the amazing ecosystem and species we encountered.

While at the camp we saw whales, Arctic fox, caribou, and birds. Lots of birds. We were smack in the middle of 23,000 pairs of snow geese, and saw dozens of other species of Arctic birds: ducks, terns, gulls, raptors, loons, and partridge. There were savanna and tree sparrows, Lapland longspur—we were proud of the additions to the trip bird list we kept. It was nesting season for all, and willow ptarmigan would land on top of our bunkhouse and wake us each morning with their ridiculous clucking and harrumphing calls. Even though it was light nearly 24 hours a day, somehow the birds knew it was morning.

Although it was June, in the Arctic it was early in the season, and we didn't get to see a polar bear. They were still on the ocean ice hunting seals; nonetheless, we were required to carry a gun for our protection, and our complex was surrounded by a single-strand electric fence to ward them off. There was a trap door in the roof of our bunkhouse, and each morning the first person who wanted to leave the building had to go up and make sure no bears returning early from the sea ice had broken through the fence in the night. We were a bit nervous about polar bears, and when we heard that a female and cub arrived at camp the day after we left, we didn't know whether to be glad or disappointed.

Our five comrades were botanists, biologists, and graduate students studying the relationships between tundra vegetation and various bird populations. Rocky assigned Zach, Terry, and me 200 snow goose nests each, which we monitored daily. The study sites were a few miles across the tundra, so we struggled across the soggy landscape in hip boots following compass bearings because everything on the tundra looked alike. In 1995, we did not have GPS and had to follow maps carefully. The landscape was so similar that, at first, we could hardly see any landmarks. By the end of our month, we had become a bit more attuned and could recognize a particularly tall shrub or deep pool. If you got off course and missed the study site, you had to backtrack and carefully recalibrate as the small flags we marked each nest with only showed up at close range. At data points we numbered the nests and recorded egg production, predation, hatch information, and nest and habitat characteristics. We walked miles of tundra each day, leaving camp in the morning and returning exhausted at suppertime. Several times we saw tall animals on the horizon. An immediate adrenaline rush was usually followed by relief that it was just a caribou, not a bear. Anything as tall as three-foot looked huge on that flat, uniform landscape.

Rocky's project centered on the impact of lesser snow geese on Canada's fragile Arctic tundra, whose over-grazing was causing potentially permanent damage. The population had steadily grown over the last two decades, and the snow geese were crowding out the Canada geese. The botanist crew had built snow goose exclosures to document the difference made by the snow goose's grazing. Inside the exclosure, the vegetation was lush and green and diverse. Outside, it was sparse and patchy and dominated by mosses. The snow geese, it turned out, use their more serrated bills to pull up the vegetation by the roots, killing it. Canada geese on the other hand are grazers, merely clipping off the vegetation, which allows it to recover.

Rocky's data suggested that changing agriculture practices in the southern United States, where snow geese winter, were affecting the population. In the past, livestock had been let into the grain fields after the harvest to graze, but now many U.S. farmers had moved livestock inside, into concentrated animal feeding operations (CAFOs). The unused grain was thus available to the snow geese. This surplus of food for the geese made them so healthy that when they returned to the Arctic, they were able to lay more eggs and hatch more goslings than the habitat could support. They were crowding out the Canada geese, whose grazing style was much less destructive to the fragile tundra.

Rocky's long-term study documented this, and his answer was longer hunting seasons for snow geese. He even contemplated commercial harvest of snow geese. Now, many states in the U.S. have more liberal bag limits for snow geese, and North and South Dakota, Kansas, and Nebraska have added spring hunts.

During our month, the last of the winter snow disappeared and spiky coltsfoot flowers popped up from the tundra. Labrador tea and the other tundra shrubs also flowered. The weather warmed enough that we even got hot on our tramps home to the station. Along with this heat, the mosquitos hatched, and on a windy day there would be a solid cloud following us, using our bodies as windbreaks. On days with no wind, we suffered and covered up with long sleeves and bug spray. Before we knew it the summer solstice was past, and with it, the days that were so long we had to put cardboard over the hut windows so we could sleep in the 24-

hour daylight. Just as we mastered the navigation and bird identification, it was time for us to leave and the tundra buggy came chugging over the horizon with supplies for those staying on for the entire summer.

We packed up and left with mixed feelings. We had learned to love the bleak landscape and calls of the geese all around us, but we would not miss the summer's slew of biting insects. When we got to Churchill, it seemed a veritable metropolis, noisy and crowded. We turned our minds to the next leg of our journey which would traverse the Yukon, Alaska, New Zealand, Australia, Thailand, and Bali. Those adventures are for another time and place but took us through the summer and fall.

A View From the Top

It was November of 1995 when we arrived in Nepal, where we celebrated Zach's 19th birthday on top of a mountain in the Himalayas. The birthday celebration was at a tea house in a tiny village next to a Buddhist monastery. The settlement was built on a small plateau surrounded by peaks. It had a few cultivated gardens and prayer wheels along the walking paths. It was a beautiful but rugged place to live. The women collected firewood in huge bundles on their backs for cooking, and we saw them washing their long, dark hair in icy-cold water from the snowmelt stream that ran along the valley floor. We stayed in the only tea house, which was owned by a Sherpa who had summited Mount Everest three times, leading two Japanese teams and an Italian one. I asked him about the jubilation on reaching the top of Everest, and he retorted that it felt terrible, he hated it, and no sane Sherpa would ever climb that mountain if it weren't for the money. He went on with some gusto to describe how the human body starts to die at that altitude. We crazy Westerners, he said, were the only ones who thought it was great up there.

Our time in Nepal started with several days in the capital, Kathmandu, which is a perfect place to understand the true meaning of culture shock. One day, shortly after we arrived in the city, I was sitting on a busy downtown street corner, taking a break from our trip preparations, drinking a cup of coffee. I could hear Bob Dylan's tune "Tangled Up in Blue" drifting out of a nearby café. As

I watched, a Rolls-Royce drove by, and a minute later I saw a herd of cows coming down the middle of the street. The bovines took a leisurely stroll past me, and one of them defecated on the street. That cow pie had barely hit the pavement when an old woman ran out into the street and gathered it up in her bare hands. As I watched in puzzlement, she reared back like a baseball pitcher and heaved it onto the side of a nearby building. It was then that I realized that the brown circles on that wall were not decorations but cow pies in various stages of drying. As the pies dried, they fell from the wall, and the old woman would use them as stove fuel to cook her supper. Within the span of an hour, I was approached by prostitutes, begging lepers, and a legless man on a cart with chair rollers who wanted to sell me street food. Kathmandu was a far cry from Mercer, Wisconsin.

We got our biological work in Nepal through a connection of Terry's. In 1982, she worked on a project in the Apostle Islands National Lakeshore looking for ways to reduce conflicts between birds and commercial fishermen. The double-crested cormorant was a state-listed threatened species at the time, but their numbers were growing, and they had taken to foraging inside the commercial fishermen's pond nets. These nets were a clever trap for fish, with an underwater "fence" that led fish to a bowl-shaped net "pot" where they would mill around until the fisherman arrived to lift the net. The problem was that pot was held in place by wooden poles that provided a perfect perch for cormorants to roost and dry their wings. Of course, the cormorants also saw the fish in the pot, an irresistible temptation for any fish-eating bird, so they would dive in and forage damaging a lot of fish they didn't even eat. Needless to say, the fishermen were not happy. The lead biologist on the project was a woman named Esther Lev, and after the cormorant work was done, her adventurous career had taken her to India, Nepal, and Thailand, where she studied birds and led tour groups. When Ester heard that Terry was traveling the world and looking for some wildlife experience, she suggested Nepal, where her good friend, Hem Sagar Baral, was one of the leading ornithologists.

We were lucky to find Hem in Kathmandu, as he worked for an eco-tourism company and was often gone on tours. He had just returned from fieldwork in Langtang National Park high in the Himalaya Mountains and was soon to depart for the Netherlands. We met him for dinner and after some discussion he said he could put us to work. He was studying fork-tails, a group of small thrush-like birds that inhabit stream and river valleys in the mountains. He was working with a man named Sebastian Buckton, a research assistant at the University of Wales in the United Kingdom, who was studying biological diversity of river birds in Nepal. Sebastian was in the field, and Hem suggested we might be able to find him on our trek and assist with mist-netting.

"There is one problem," Hem said. "Sebastian is still up there in Langtang completing the winter survey work and we don't have communication. There is no sure way to find him. I expect he might be along the main trail between our study locations, but Langtang is a big place."

We thought it over. What the heck, we were here and even if we didn't find him, we had been thinking about hiking the Langtang trekking route anyway. It was less touristy, and Esther had told Terry it was beautiful. The next day Hem handed us maps and a sealed envelope to give "Seb" if we found him. Before we parted I asked, "By the way what does Sebastian look like?"

"Oh, you'll know if you find him," Hem said. "He's British, blond, and tall. Don't forget to give him my letter."

We left the next day on foot, taking it slow the first few days to adjust to the altitude. At higher elevations the air thinned and oxygen decreased, causing a few headaches but we trekked on, navigating the winding rocky trails. After a day or two the headaches went away, along with the sore muscles from our heavy packs. One day we saw a Himalayan black bear across a valley. As we continued we saw several species of monkeys, and even a red panda, which ran across a mountainside within view and disappeared into the high forest. There were other travelers from around the world who shared our simple tea house meals and rustic lodging, hosted each night by different Nepalese families.

One day, after about a week in the mountains, we were walking up a steep stretch of trail and rounding a bend, when we saw a couple of porters followed by a guy in khaki shirt and trousers. He had blond curly hair and a mustache, looked about 30 years old, and very British. The real clincher was that he wore a pair of expensive-looking binoculars around his neck. The scene reminded me of what it must have been like in Africa when Stanley met Livingston. We approached and asked, "Are you by any chance Sebastian Buckton?" He looked at us with surprise, and then in a strong British accent replied, "As a matter of fact, I am." We handed him our letter of introduction, a bit worse for wear after a week in my backpack. Seb opened the letter and carefully read it. He then slowly looked up smiling and said, "Welcome. Fall in line. We'll be stopping for the night at a tea house about two kilometers from here."

We went to work for Sebastian, mist-netting and banding Himalayan river birds. A mist net is a fine mesh nylon net stretched between two tall poles like a fence. When set up, the net is nearly invisible. Unsuspecting birds that fly into it become tangled in its pockets allowing a researcher to safely extract them for banding. Sometimes a couple of us would run the nets and the other two would go upstream and try to "herd" birds down to the net. We worked locations at different elevations starting at dawn each day. We caught spotted and slaty-backed fork-tails, brown dippers, river chats, and plumbeous water redstarts. These cold, cascading streams were misty and cool. The birds were secretive, flitting downstream in the mist. Eventually, we worked our way down the mountains and back to Kathmandu. Had we skipped this research work we would never have experienced the magic of these valleys in the early morning light. The scientific work made us slow down, pay attention to details, and appreciate the complexity of the river systems. And of course the dominating presence of the snow-capped mountain peaks rising around us. Culturally and scientifically, Nepal was one of the highlights of our trip.

I can still picture one small white stupa, a form of Buddhist monument, that stood high on a mountainside. We had hiked toward it for a couple of hours, and when the switchback trail finally approached, we could see a calm statue of Buddha sitting inside. There were prayer flags flying from its pinnacle and it was surrounded by snow. Everything was quiet and beautiful, and it seemed like a sacred place. That stupa made a very strong impression on Terry, and she later made a large oil painting of that mountain view, showing herself on the trail below, dwarfed by the calm panorama. Nepal taught us much more than what birds live in river valleys. It was a bit of a pilgrimage, testing our strength and showing us the grace and kindness of the Nepalese people.

Six Degrees of Separation

In January 1996, we arrived in Guatemala, where we'd arranged to meet up with Terry's sister Diane at Tikal National Park, in the northeastern Peten region. Tikal is one of the most significant Mayan archeological sites in Central America and certainly the most famous in Guatemala. From 700 B.C. to 1200 A.D. it was home to a population of 100,000 people, and according to a United Nations website, "Tikal National Park is one of the few World Heritage properties inscribed according to both natural and cultural criteria for its extraordinary biodiversity and archaeological importance." The archeological site is an ancient city complex of around 3,000 temples, ball courts, pyramids, palaces, and structures that covered 130 square kilometers. Today, Tikal is a United Nations designated Biosphere Reserve and is also the largest intact tropical rainforest in northern Central America, covering 17,000 square kilometers. Within Tikal's overgrown ruins live over 300 species of birds and 54 species of mammals.

The Tikal Reserve has some modest accommodations for tourists, including a small hotel and café, and a rustic campground. Diane had arrived a couple of days ahead of us to scope out the camping situation and set up her tent and hammock. This was great, because after greeting Diane, we just followed her suggestions on where to set our tents, how to find the restrooms, get showers, firewood, and food. You name it, she had it all figured out.

After we'd set up camp, we started talking. Diane said, "Jeff, I saw something interesting you're going to like. Right at dusk last night a strange raccoon-size animal walked out of the forest and right past me before heading back into the jungle—it had a bright red numbered tag attached to its ear." My curiosity was piqued. I thought the animal was probably a coatimundi, but I'd never seen one before, and the ear tag meant there was some research going on.

I asked the Guatemalan campground attendant if he knew anything about the ear-tagged coati. My Spanish was not too good, and his English was not much better, but eventually he understood my question and pointed to a large canvas tent in the corner of the campground. I didn't want to appear too nosy, so I did a casual walk by and peered through the netting. I was quite surprised to see a stack of Tomahawk cage traps sitting in the screened porch of the tent. This seemed like a crazy coincidence as Tomahawk is a trap company from northern Wisconsin, and I use their cage traps to capture skunks and raccoons.

I kept an eye on that tent and, sure enough, by the late afternoon a man and a woman walked out of the jungle and went into the tent. I waited a few minutes until they came out, then approached, introducing myself, and asked if they were studying the coatis. I shared that I worked as a wildlife technician in Wisconsin. Yes, they said, this was a PhD project, and they were from the University of Florida in Gainesville. They were familiar with Wisconsin because they had both done their master's degree work at the University of Minnesota–Twin Cities. I immediately asked if they knew Mark Martell, a friend of mine who worked at the raptor center affiliated with the university. They smiled and said Mark was a friend of theirs too. I expressed my interest in their work, and we agreed to meet at the small concession restaurant for dinner that night. "Wow," I thought, "what a small world."

After the six of us got seated we went around the table introducing ourselves. The coati researchers were Gerry Binczik and Sue Booth-Binczik. They talked a little about their background, their master's work in Minnesota, and their current project in Tikal. Next was Terry, who described her work for an environmental institute at a small college in northern Wisconsin that she had attended as an undergraduate. Gerry immediately stopped her. "That's interesting," he said. "I did my undergraduate degree at a small college in northern Wisconsin too. Where did you go, Terry?"

"Northland College in Ashland," Terry said.

"That's where I went! What year did you graduate?"

"1981."

Gerry said. "That's the year I graduated!" Now Gerry and Terry really started staring at each other. You see, Northland College is not that big of a place, and in 1981, total enrollment was about 850 students.

Terry asked, "What dorm were you in freshman year?"

"Fenenga Hall."

"I was in Fenenga too." The next question was what floor, and yes, they'd been on the same floor in the co-ed dorm. Finally, Terry and Gerry remembered each other. They had lived just down the hall from each other, but back in those college days Gerry had been quite a party guy, with hair down to his shoulders. He looked a lot different today. Terry, for her part, had been shy and socialized with the granola, tree-hugger kind of crowd.

As we continued around the table with the introductions, it came out that Diane and I also attended Northland, and Zach was enrolled to attend there that fall. The world was getting smaller, like the proverbial "six degrees of separation" that says that, on

average, all people on the planet are connected by fewer than six social connections.

The next day, we joined Sue and Gerry on their coatimundi study. The coati females roamed in bands of 50 to 60 animals, with perhaps one male in the group. We would climb the ancient Mayan pyramids to locate the radio-telemetry signals from their collared animals and then follow them from the ground. The coatis would allow us to approach, and suddenly we would be surrounded by dozens of the animals scurrying everywhere. We could watch them hunt and feed and observe their social behavior. To capture one for collaring, Gerry would deliver the drug by blow dart, using a homemade device with a pressurized syringe tube. On impact, a little rubber stopper would slide up and deliver the drug. The study was looking at the fertility of the males, so we inserted a tiny tube and collected a sperm sample from them. Using our high-tech equipment amid Tikal's ancient ruins, I wondered: what might the ancient Mayans have thought of our work?

We also had a chance to experience the magic of those forests, hearing howler monkeys roaring like lions at dusk and dawn and watching sunset from the top of a huge pyramid in the "Mundo Perdido," or lost world, an area of Tikal that seems particularly remote and mysterious, especially at dusk. There were brightly colored tropical birds flitting through the dense forest and the sounds of unfamiliar insects. Diane and Terry will still fall into fits of laughter remembering an encounter I had with the endangered oscillated turkey. This good-sized bird looks like our wild turkey but is more colorful. We had been warned to steer clear of them as they could be aggressive. In particular, we were told not to run if challenged. Well, I couldn't resist approaching a group of turkeys for a better look, discounting the warnings as a bit over-the top. Before I knew it, there was a large male heading my way, and I tried to fend him off with my daypack like a matador with a cape. Before I knew it, I was running, followed by a whole flock of irate turkeys, which served to encourage my avian adversaries even further. Needless to say, I survived the encounter but my pride was taken down a notch or two.

Joking aside, Tikal presents moments of wonder at every turn, from the lines of leaf-cutter ants who harvest small leaf parts to feed their underground fungus gardens, to the three-foot-long hanging nests of the oropendula birds. On a more esoteric level, Tikal is a vivid place to ponder the collapse of human civilizations and our own fragile existence on the Earth. Scholars suggest that the Mayan civilization failed due to overpopulation, environmental degradation, warfare, extended drought and other factors. Today, as we contemplate global issues such as world hunger, climate change, and ocean pollution, we would be wise to take a lesson from the Tikal story.

After our year of travel, we returned to Mercer with these and other stories to share. But we were not the only people having adventures that year. When I started back at my job in Mercer and met Claire Gower from England, I got to hear stories of her adventures traveling in "foreign" places. You might recall from the beginning of this long story that I began by talking about Claire, the gal who blew into Mercer on the wind from the "Butterfly Effect." It turned out that she had been a real hit at the ranger station. She had reorganized the office, done all my work perfectly, compiled some old survey data of Bruce Bacon's, and co-published it with him. She and Steve Kangas had done such good work, I felt lucky they gave me my old job back.

The DNR crew had done their job on Claire too. They'd found her living quarters with one of the foresters, and turned her into a deer hunter, grouse hunter, and angler. She learned how to run chainsaws, band birds, and drive boats, tractors, ATVs, and snowmobiles. The work taught her animal track identification and surveys for winter furbearers, frogs, ruffed grouse, woodcock and wolf packs. Claire also ran nets with the fish crew and fought wildfires with the fire crew.

Bruce took her to some of our state wildlife meetings and research conferences, and she got to know a lot of other biologists, including Bill Mitten, our state big-game biologist. Bill had some connections out west in Montana and after Claire's stint here at Mercer, Bill found her two months' work near Glacier National Park with the Blackfeet Nation Fish and Wildlife Department studying grizzly bears. To make a long story short, Claire went back to Europe where she finished her undergraduate and master's, but she found that she longed to return to the States. She started seeking opportunities, and after some perseverance, got a job studying predator-prey relationships between wolves, elk, and bison in Yellowstone National Park. This would become her PhD work. Today she is a waterfowl, nongame, and furbearer biologist with the Montana Department of Fish, Wildlife and Parks.

That year's travels changed our perspectives. We learned about other cultures, other ecosystems, and came home with new ideas about our life here in the U.S. We saw firsthand how actions in one part of the world can change life halfway around the globe. The intertwining threads of our adventures helped shift the trajectory of Claire Gower's life, as our futures were changed by encounters with Rocky Rockwell and Sebastian Buckton. We arrived home changed by countless seemingly random events. It was as if we had lived a bit of Lorenz's chaos theory and to this day I believe that if a butterfly's wing can change the world, we can too.

Innocents Abroad

"We are off searching for the meaning of life… we will be back when we find it… or get hungry." —*Message posted on Jeff's and Terry's cabin door, February 2019*

The Ivory Hunters

Terry and I were quite excited when, in 1999, the Cornell Lab of Ornithology made the announcement that the ivory-billed woodpecker had been rediscovered. Could this really be true? The "ivory-billed" had been declared extinct decades ago. It was an impressive, large woodpecker sometimes referred to as the "Lord God Bird." The last bird, a lone female, had been observed in 1944.

Because the announcement came from the Cornell Lab and the United States Fish and Wildlife Service, most biologists assumed there was firm evidence behind the claims. According to Cornell, the discovery site was the White River Wildlife Refuge in Arkansas. But it was a puzzle—where had this extirpated species been hiding since the last sighting in 1944? Its picture had even been removed from some modern field guides, and its sad story was a well-known cautionary tale about the delicate relationship between people and animals.

The species' last remaining population had been discovered back in 1935 by an ornithologist, also at Cornell Lab, named Arthur Allen. A student on the discovery team, James Tanner, took up the cause and did a research study of the birds, documenting their decline to extinction. No birds had been seen since. The tragedy was that the birds could possibly have been saved had their habitat been protected. Tanner discovered that the species needed large tracts of river bottomland, forested with hardwoods that were large and included a mix of dead trees. That last dwindling population was living in Louisiana in a large block of bottomland forest owned by the Singer Sewing Machine Company. Despite pleas from the scientific community to protect the habitat, Singer refused and logged off the tract in 1941. It was wartime, and the wood, they said, was needed to build cabinets for sewing machines used to sew uniforms and tents for the troops. Stephen Lyn Bales' book, *The Grail Bird: The Rediscovery of the Ivory-Billed Woodpecker*, about Tanner and the demise of the ivory-billed cites a bold quote from the chairman of the Chicago Mill and Lumber Company who logged the Singer Tract, "We are just money grabbers. We are not concerned, as you folks are, with ethical considerations." Public attitudes have come a long way since then. Back in the 1930s we had no Endangered Species Act, and never would have considered changing forest practices to save a species as we've done recently for the spotted owl or Kirtland's warbler.

It wasn't long after Cornell's announcement that Terry and I read an article in the magazine *Florida Birder*. The article stated that an ornithological team from Auburn University in Alabama had recently discovered another possible sighting of ivory-billed woodpeckers in the panhandle of Florida. Interesting, we thought. We traveled to Florida each winter to visit Terry's parents—why not contact Auburn University and see if they could use some volunteers? We sent our resumes to Dr. Geoff Hill at Auburn and heard back from him right away: Yes, they could use our help.

Dr. Hill noted my last two years at Wisconsin DNR as a forestry technician, and asked if we could inventory forest plots throughout the bottomland hardwood to contribute to a habitat suitability index. This would help to better define the habitat requirements of the ivory-billed should the sightings be confirmed. Dr. Hill had a graduate student and research team headquartered in a rented cabin near the study site in the Choctawhatchee River basin. The forestry data would also be used to distinguish other areas where ivory-billed woodpeckers could be released if birds were located and captive rearing ever attempted.

We knew absolutely nothing about southern bottomland hardwood forests. We didn't even know what tree species would be growing within our inventory plots. We had spent our entire careers in temperate and boreal forests in the Upper Great Lakes region. "Don't worry," Dr. Hill said. "My graduate student will train you."

We told him we could give him eight days, packed the car with anything we thought we might need—canoe, tent, camping gear, binoculars, spotting scope, chest waders, GPS unit, bird books, mosquito repellent—and proceeded south.

The evidence supporting the re-discovery of this extinct bird was quite compelling. Several reputable ornithologists at both sites in Arkansas and Florida claimed to have seen the bird. There were also recordings thought to be the bird's "kent" call and its double-knock drumming call, and measurements of possible nest holes. The data from Tanner's 1935 study in Louisiana had noted that ivory-billed nest cavities were wider and more elongated than those of their close cousin, the pileated woodpecker, which can also be found in similar habitat and has overlapping range. The earlier study also found that ivory-bills often left sign of peeled bark they had pried away with their strong bills to reach beetle grubs, and the way they did the prying could be distinguished from that of pileated woodpeckers. The researchers felt they had found some evidence at both sites, but it was only circumstantial. No one had a clear photo or feather for DNA to prove beyond a doubt that the species still existed. It was also problematic that ivory-billed woodpeckers share many physical traits with the pileated. The species are of similar size, sport a red tuft on the head, and have a black tail and black feet. The ivory-billed

ivory-billed

was slightly larger, had a cream-colored bill and a skunk-like white stripe down its back. If observed flying at much of a distance, even expert birders would have a hard time distinguishing between the two species. Terry and I thought, wow, what if we were the ones to get the perfect picture or find a feather or discover an active cavity roost?

When we arrived in Ponce de Leon, Florida, a tiny town in the middle of nowhere, we met Brian Rolek, the graduate student we"d be working for. Brian had been on that first expedition with Dr. Hill when they had initially observed what they believed was their ivory-billed. There were two other team members at camp, a male undergraduate and a woman who was a contract biologist and had worked previously at the Cornell site in Arkansas. She had some great cottonmouth (water moccasin) snake stories to share. They had tried to crawl into her canoe numerous times, had crawled onto stumps next to her and gone to sleep, and she had even observed them copulating. The good news was that all these encounters occurred in warm weather, and at the moment Florida's panhandle was pretty cool, making the venomous, cold-blooded snakes sleepy.

The next day we received our training to conduct habitat site analysis. The bottomland hardwood forest in our study area was a mixture of tupelo, oak, spruce, cypress, sweet gum, and pine with an understory of palmetto. It was common to have trees over four feet in diameter, and on the cypress trees their knees (the cone shaped structures on their roots) were up to six feet tall. Terry and I were in another world. We had a grid of GPS locations to visit and habitat plots to measure at each site. We were really going to experience the backwoods of Florida up close and personal. We headed out that afternoon, up the Choctawhatchee River by canoe, with food and supplies for eight days. Our plan was to set up camp near the center of our plots so we wouldn't have to move camp very often. Brian had said, "Try to locate your camp on high ground, because two inches of rain upstream in Alabama equates to a two-foot river rise down here."

We paddled several miles the first day, found a nice site, and set up camp. The next day we set to work, all the while marveling at the cathedral of huge trees, braided with rivulets and creeks. The method was to canoe as close as possible to our plot centers, then tie the canoe and carefully slosh through the watery habitat to the exact GPS location. Care had to be taken as the water was dark and deep holes or tree roots were not visible. Water filling our chest waders would have made for a very uncomfortable day. We recorded the plot data in about 20 minutes, then moved on to the next, usually half-a-mile or more away. Tree species, tree density, height, diameter, ground cover, and presence and absence of snags and tree cavities, were all noted using many of the same instruments I used in Wisconsin while conducting forestry reconnaissance. During our work we saw beaver, deer, wild boar, and squirrels and recorded pileated woodpeckers, hawks, owls, and numerous songbirds, but we saw no ivory-bills.

Then it started to rain. It was our third day, and it began in the late morning and continued all day and into the night. It was a hard rain, and who knew how much had fallen further up the watershed.

It was Terry who saved our camp. She has always bitterly complained that I can sleep through any tempest, leaving her stuck with the nocturnal checks for loosened tent stakes, drifting boat anchors, or bears sniffing out a midnight snack. She woke up in the middle of the night and decided to look outside the tent to check the river. She didn't have to look far, because the river was about two inches away and rising. She called the alarm and we put on our head lamps and started packing. In 10 minutes we had everything in the canoe and were paddling for higher ground. We reset camp and got a good night's sleep, but Brian's warning had been accurate.

We completed our work on schedule and spent the last night back at the research cabin with the rest of the crew. They had been installing sound recording equipment to try to record the suspected ivory-billed kent calls and double-knock drum call. They had also been searching for tree cavities to measure for possible nest or resting sites. They always had their cameras ready in hopes of a possible photo, but there had been no sightings while we were there.

Dr. Geoff Hill eventually wrote a book on the ivory-billed woodpecker titled *Ivorybill Hunters: The Search for Proof in a Flooded Wilderness*. Brian Rolek completed his master's thesis, yet no definitive proof ever surfaced that the ivory-billed had truly been found. A year or so later, the Cornell Lab of Ornithology released a statement that their observation and few seconds of video were not fully substantiated, and they abandoned the search for the elusive ivory-billed. Much was made of this admission and monies expended on the fruitless search for the "Lord God Bird."

Terry and I felt some empathy for the biologists who had worked on the search. We talked about how easy it would have been to wrongly identify a pileated when we were in that bottomland forest. We saw quite a few and each time our adrenaline rose while we scoped the birds for telltale markings. The hard part was that they were usually on the wing and disappearing before we got our feet on stable ground and could raise and focus the binoculars for a clear look. It was a magical forest, with its giant trees, knobby cypress roots and the continual rustle of the palmetto fronds. It was also slightly eerie when rising waters erased all landforms, creating a maze that could only be navigated by GPS. I can still feel the quick surge of excitement we felt when we saw the dipping flight of a pileated in that Choctawhatchee bottomland. We were saddened that the ivory-billed was again declared extinct, but after our adventure we came away with much deeper appreciation of what that king of woodpeckers must have been like.

A Cowboy Named Hank

Terry and I met Hank in Southern Colorado, way up in the San Juan Mountains during an early fall snowstorm. But the story begins with a great friend named John Depue, who was a summer employee at the Mercer ranger station. John had a really hard job, working with me tramping around the Iron County Forest chain-

sawing brush and trees that invaded the grassy wildlife openings the DNR maintained. On an average day, we went through a gallon of water each and half a bottle of mosquito repellent. John was a trooper through it all, and we became good friends, staying in touch after he moved on to new challenges. John was about my son Zach's age, and after he left Mercer, he got his master's degree studying river otter in Idaho. He designed a technique for collecting wildlife DNA which uses a small wire brush to snare otter hair. His novel method is used today by many researchers.

John contacted us in 2004 to see if Terry and I could go to Colorado for a month to fill in for him on a wildlife project. He had been hired by Tanya Shenk, the biologist with Colorado Parks and Wildlife in charge of the state's Canada lynx reintroduction project. Lynx had been extirpated from Colorado for many years and the state was attempting to re-establish them by acquiring lynx from Canada and Alaska where the population was high and releasing them in Colorado. The cats were tracked using radio collars. John's job was to recapture the female lynx and replace the fading collar batteries. So far, he'd caught five out of 16.

I talked with John for a while before he confessed the real reason he needed us. He had been on this lynx project for a year with no time off and was burning out. He and his girlfriend wanted to go mountain climbing in Greece for a month. John is adventurous, and his idea of relaxing is kayaking a whitewater river or climbing a mountain. He had asked Tanya for a month off in September, but she had said no; her priority was to re-collar the lynx. Lynx populations in Canada and Alaska are very cyclic, and Tanya wanted to assess whether the re-introduced animals were producing young, which involved following them to their natal dens using telemetry. The project's success depended on determining if the females were having kits and whether those kits were surviving. It's common for lynx to commit infanticide if they're not healthy enough to feed their young.

After her initial refusal to grant his leave, John asked Tanya if she would re-consider if he found a qualified replacement to cover the month. Cautiously, she said yes. He was on the spot to find someone who could come on short notice, had trapping experience and drug-handling certification, knew telemetry equipment, and would be willing to camp in the mountains. Tanya had reminded John that lynx were an endangered species in Colorado, so qualifications had to be water-tight. John said, "I know just the guy."

Terry was enthusiastic, saying, "Let's go. It'll be like a paid vacation." When the approval came through, we loaded up the foot traps, ordered lynx lure and lynx urine from a trapper supply company, packed the truck with camping equipment, and headed west.

When we got to Fort Collins, Tanya gave us some bad news. A couple of days prior to our arrival, their pilot, the air support to help us find the radioed lynx, had crashed and died. There was an investigation into the accident, and until that investigation was complete all pilots were grounded. She told us that the country was too big to go looking for lynx without air support. The terrain and size of the cats' home ranges made it more efficient for planes to locate the lynx from the air and provide a nearby GPS location. That allowed the trapping crew to head in on foot with a handheld antenna and set traps as close to the lynx as possible without spooking the animals.

A week later Tanya called to tell us the planes were back up and one had found a cat for us to work on. She gave us the location near Telluride, and we headed out. We soon learned that telemetry was very difficult in the mountains, because radio signals bounce and echo off the slopes. We would get a strong signal and head for the source, only to find that it was now coming from a totally different location. As flatlanders from the Midwest, this was new to us. We finally figured it out enough to locate the first cat and start setting traps.

John's technique was to sneak to within a quarter mile of the cat and set lots of traps. We made baited dirt-hole and cubby sets using our lynx lure and urine as an attraction scent. We also added an "attractor" of feathers or fur above the set for eye appeal. This was important because the lynx is a much more visual than olfactory predator. Imagine trying to attract your pet kitty to play with a nice moving toy—these were the types of sets John had been successful with. We were able to make eight sets before it got dark and then retreated and set up camp.

Sometime after dark it started to snow and by morning our tent was heavy, laden with wet snow. Even though it was only September, at higher elevations snow can come early. This was problematic as the protocols for the project stated no foot trapping was allowed in snow to prevent the risk of a cat freezing to death. In the morning, we pulled our traps, drove down to where we could get a cell phone signal and called Tanya.

She endorsed our decision, saying that with a high-profile project where there was lots of news coverage, she would rather not risk injuring a cat. She went on to explain that many people were not pleased with the idea of re-introducing another predator, "Remember this is the West, and it's sheep and cattle country. Some ranchers hate predators, so don't ever tell anyone what you're doing." The forecast for Telluride called for more snow, so Tanya suggested we shift operations to a location further south where the chance of snow was less. The pilot had found another cat about 10 miles from the New Mexico border.

The second cat was high up in the San Juan National Forest. We followed the signal to the end of a dirt forest service road where a pickup and empty horse trailer were parked at a trailhead. We decided to hike up the trail a bit to try to pin down the cat's location, and while we were unloading our gear, a woman in a Jeep Cherokee pulled up. She was in her 50s, attractive and well-dressed, and I thought it odd that she would be out here in the middle of nowhere on this narrow mountain road. She rolled down her window and asked us if we'd seen Hank.

"No," we said. "We don't know any Hank. We're not from around here. We're just camping."

She told us the trailer belonged to Hank, and she was worried about him, "He's 83 years old and has had two heart attacks. He's been up there somewhere all alone on horseback for several days and was due back yesterday. If you see him, tell him to get back to the lodge in Platora. He's in big trouble with me." We asked if she was a relative. "No," she said, "just a friend, tell him Jane is looking for him."

We took off after the lynx. When we thought we had found it in the next valley, we set traps, then pitched camp. That night to our dismay, a wet, heavy snow fell. Our two-man tent was practically flat under the weight, and in the morning, the snow was eight inches deep. We pulled our traps and headed for our truck, thinking that this "vacation" was not all it was cracked up to be. We were cold and a little let down. When we got to the trailhead, Hank's truck and horse trailer were gone. There were no fresh tracks so we thought he must have come in before the snow.

We considered our options. It was clear we couldn't stay there, as we would barely get down the steep road in the snow, even with four-wheel drive. We also needed to call Tanya. Terry said Platoro, that little hamlet we'd passed yesterday about 10 miles down the road, was our only hope for a phone.

About an hour later we pulled into town. It was an old mining town that had turned into a kind of ghost town. Everything was closed up except for what appeared to be a hunting lodge. The rest of town consisted of old miners' houses converted to seasonal cabins, empty this late in the season. We drove up to the lodge where a couple of cars and trucks were parked. A sign said the lodge took guests in the summer for trout fishing and hunters in the fall for elk. We pulled open the door and saw a big room with a huge fireplace and, best of all, a blazing fire. The fireplace was so big, you could stand up in it. There was a cowboy and a couple of women standing around, including Jane. A cowboy looked over at us and said, "Come on in. We've been wondering how long it was going to take before you showed up." He handed Terry and me each a steaming cup of coffee, asking if we needed cream or sugar.

"Both, thanks," Terry said.

The cowboy said, "Don't tell us some tall tale about what you're doing here. We know you're following that damn lynx. I checked out your truck at the trailhead and saw your equipment. I've seen those lynx tracks up there for two years and seen it a couple of times on the trail. Now, don't get all worked up about that cat. We like it and I'm all for the state releasing them up here. If I didn't like it, I'd have already shot it. Relax, take off those wet coats. Come by the fire and warm up." We'd met Hank.

It turned out the lodge had been bought about a year ago by Pam, a lawyer from California. Pam knew little about running a hunting and fishing outfitter or lodge. Hank had worked for the previous owner, so Pam kept him on to run the outfitting program at the lodge. Pam would have been lost without him.

Now, Jane's story was different. She had met Hank years ago on a fly-fishing trip where he guided a group for trout at a lake high in the mountains. They had packed in on horseback, and she'd become infatuated with Hank in a father-daughter sort of way. Jane was married to a doctor in Denver, but every chance she could she visited the lodge just to hang out with Hank. He wore a cowboy hat, jeans, and well-worn chaps. His cowboy boots had spurs on them. What really captivated me was the way he talked—real slow—so that when he said something you wanted to listen. Terry and I were immediately taken with him, just as Jane had been.

We called Tanya to get advice on the cat, and she told us to stay put. The forecast suggested the snow would melt in a couple of days, and she had no other cats lined up with decent access.

"Where are you?" she asked. We told her where the lodge was. "Good, stay there, but don't tell them what you're doing." We told her it was too late. These people knew more about this lynx than we did. But not to worry—they supported the project.

So, we settled into a room and had a break from hard work. The lodge wasn't crowded, but they had people coming in soon—elk season opened the next weekend. Each day, waiting for conditions to improve, we assessed the snow and tried for some telemetry readings on our cat.

A day or so later Hank asked, "Can you two ride horses?" Terry said she used to own a horse. I told him I grew up with horses on the farm back in Iowa. "Well," Hank said, "I've got to go up in the mountains and set up a hunting camp. You know, big canvas tents, folding stoves, grain for the horses' food, that kind of stuff. We'll haul the gear on pack horses. Would you mind helping out?" We'd love to, we said, and headed out the next morning.

We really got to know Hank and appreciate his skills in packing up a horse, riding, and most of all storytelling. He had an endless supply of stories, not bragging stories, but life adventures. We now knew why Jane spent so much time with him. He told us about the mountains and cowboy life, about past elk hunts. When he spoke, we were all ears.

Late in the day when we returned to the lodge, Pam told Hank some bad news. She had paid her kitchen cook and waiter their monthly wages and taken them to town to do some shopping. They were a married couple. Apparently, they jumped on a bus and skipped town. Hank knew what had happened, and he told her he suspected they'd lined up a winter job at a ski hill.

"They knew we were closing in two weeks just after elk season. Rule number one," Hank stated, "is never pay the help until the last day of their contract is up."

"What are we going to do?" Pam asked. "The hunters are already starting to show up."

"Relax," Hank said. "We'll figure something out. You and I can cook for those guys. They're not particular."

"What about the waitressing and dish washing?" Pam asked.

They slowly turned and looked at Terry and me. There was silence. Terry looked at me, then back at them. "You've been wonderful hosts," Terry said. "I used to wait on tables back in college. Jeff can do dishes. Of course, we'll help."

We never did catch a Canada lynx in Colorado. It just wasn't in the cards. We finally did get our traps out and caught quite a few coyotes, which we released. We had one lynx in a trap, but it pulled its foot out and escaped before we did the morning check. It was likely that some frost froze the trap down and it was slow to close on the animal's paw, giving it a good scare but doing no harm.

John Depue returned from Greece after a great vacation. He went back to lynx trapping and eventually caught Tanya's cats. The following year they documented females with newborn kittens. When we started back to Wisconsin, I stared at my hands on the steering wheel. They had never looked so clean.

Well-Guarded in Honduras

In early winter 2015, Terry and I were sitting in our cabin out on the Turtle-Flambeau Flowage discussing the season to come over a glass of wine. We were dreaming of adventure, having just finished packaging our venison from deer season. We had spent that fall hunting and gathering wild rice, I had finished fall trapping, and although the sunsets were great, ice fishing wasn't really producing many fish. We thought, how about heading south for a while?

This wasn't a new idea. After spending three months in Central America on the world trip, we'd returned a few times to visit countries we had missed: Costa Rica, Belize, Panama. Each trip provided vivid memories, such as the trip we took crewing on a 45-foot sailboat from Columbia to Jamaica. On that adventure we were chased back to port by 15-foot seas, sailed through starry nights taking three-hour watches, and one day stopped to swim in a perfectly calm ocean. We jumped into the cerulean blue depths, always leaving one person onboard to make sure we could scramble safely back to the deck.

Our typical routine was to fly to Florida first and visit Terry's snowbird parents, then on to some Central American country for a month, then back to Florida to decompress before heading back to Wisconsin. For some odd reason, when we vacationed this way in the deep of winter, we had a lot less firewood to make at our

cabin the following summer.

As we were discussing plans, Terry said, why not give our friend Amber a call? Amber was Dan Haskell's partner and a good friend of ours. She was teaching at Michigan Technological University where she had completed her PhD. Terry thought she might have a student working in Central America. We had first met Amber Roth when we worked for Mike Meyer studying loons. At the time, Amber was a research technician with the DNR and on the night capture team, as well as the day shift where we did nest searches and color band identification. She had done her master's thesis on golden-winged warblers, a species with a declining population of much concern to the ornithological community. This tiny bird is a neo-tropical migrant, one of many species threatened by changes in habitats both here in North America and in the tropics. Contrary to a number of declining species that need large tracts of old growth forest, or at least continual expanses of forest, golden-wingeds thrive on younger, early successional forests, like aspen clear-cuts. This bird became the poster child for paper company foresters as well as resource managers and hunters who favored ruffed grouse, deer, and woodcock, species also dependent on this habitat type.

We gave Amber a call. She said, "You say you want to volunteer to do some wildlife work in Central America? Well, why don't you go down to Honduras and help Ruth Bennett, my golden-winged warbler researcher who has a project down there. You'll love working with her. She's ex-Peace Corps, speaks perfect Spanish, and knows Honduras like the back of her hand." Amber went on to explain that Ruth was doing a master's on the golden-winged warblers' wintering ground habitat characteristics. She was up in the mountains, mist-netting and banding—doing it all alone—so she could really use some help.

We learned that Ruth was 26 years old, and I got to thinking—I'm over 60 now, could I keep up with a 26-year-old, especially in the Honduran mountains carrying a pack full of research gear? Before we headed out, Amber said we'd better check the U.S. government travel website. Ruth had indicated things had gotten a little dicey down there lately, because of the cartels shipping drugs through Honduras. All her old Peace Corps colleagues had been sent home. Ruth wasn't worried though. She was staying. In fact, the website indicated that Honduras was currently the most dangerous country for tourist travel in the world. Well, we thought, let's hope Ruth knows what she's doing.

Indeed, she did. She taught us about tropical forest ecology as well as the warblers, and, somehow, I made it up those mountain trails carrying the packs and mist net poles. We would play a tape recording of golden-winged warbler's territorial calls next to the nets, and the warblers would just fly right in looking for an intruder. We banded them with aluminum U.S. Fish and Wildlife Service bands on one leg and colored plastic bands on the other. We also documented the characteristics of the forest habitat where they were found. The birds, we found, often traveled in mixed flocks with other warbler species.

Amber was right that Ruth knew her way around Honduras. Her Peace Corps background and excellent Spanish meant she could walk into a mountain village, start talking, and find us a meal and a place to stay in no time. We surveyed and netted warblers in three Honduran national parks, and only at the last site did we have any kind of a problem. The park staff refused to let us do survey work unless we took their armed guards with us, which we did, walking along the trail with our nets, packs, and poles, while two uniformed guards walked alongside us shouldering their military M16s. We were quite a sight.

The following year, Terry and I returned to Central America, this time to Nicaragua with Amber and Dan. We captured and banded various tropical birds at a banding station on a private coffee and nature reserve. We even caught some golden-winged warblers, and Ruth came over from Honduras to join us.

Ruth Bennett eventually went on to Cornell Lab of Ornithology where she completed her PhD in ornithology. Amber and Dan moved to Bangor, Maine, where Amber is an Assistant Professor of Forest Wildlife Management at the University of Maine. Several of her graduate students are studying, you guessed it, golden-winged warblers.

A Corn Farmer's Funeral

*How rich will we be when we have converted all our forests,
our soil, our water resources, and our minerals into cash?"*

—Jay "Ding" Norwood Darling

My Dad, Del, died 10 years ago. He was 89. Mom had died a
few years earlier. I was standing at his funeral talking to some folks
when a lady from the funeral home came in and whispered in my
ear. "A car just pulled up, and it's the governor," she said. "He wants
to know what you want him to do. He wonders if you want him to
come in now or meet with the family in private. He doesn't want
to disrupt things."

"The what?"

"The governor," she repeated. "You know, Chester Culver, the
governor of Iowa."

"You don't say. Well, tell him to come on in. I haven't seen Chet
since he was a little kid, 30, maybe 35 years ago."

The governor walked in, and I gave him a big hug. So did my
brother Brad and my sister Janice. Around us, people were staring
and whispering. I couldn't hear what they were saying but I knew:
Why was the governor of Iowa at a corn farmer's funeral? It was
because of the history between my folks and Chet, and how kindly
they had treated him when he was a boy. Kindness was the cornerstone of our family philosophy.

According to my mother, during the Depression, the hobos
or tramps, as they were called back then, always stopped at my

grandfather Marshall's farm. There was a secret marking system or symbol of some kind. Mom thought it was under a nearby bridge where they stopped to get water or seek shelter from the rain, but she never found out for sure. There were a lot of people out of money and wandering the country looking for work and the symbol system was how they communicated. One symbol said, "Bad people live here, don't stop." Another warned of dangerous dogs, and others indicated kind people, farm people who would feed you and give you a place to sleep in their barn, or maybe give you some work.

Whether you were a hobo or a community leader, if you stepped into our kitchen you were welcomed, and when you sat down for some pie and coffee at the Wilson table, you heard about our concerns and views. My folks were good friends of Chet's mother and father. John Culver was a United States senator and was very concerned about rural America. Dad kept him well informed about farm life, especially during the farm crises in the 1970s when grain prices plummeted and many lost their farms. I remember Dad staying up well into the night working in his office trying to figure out how to keep our farm from going under. We had just bought a new corn picker for $60,000 and the interest rate was 16 percent.

John and Ann Culver visited us a lot when they were not in Washington D.C., and so did their kids. Chet had three sisters, and even though they were a lot younger than us kids, we took them under our wing and taught them horseback riding, creek roaming, hay baling, and the fun of farm life. On election years, my folks campaigned for Senator Culver, going door to door talking to other farmers and placing a "Vote for Culver" sign prominently along our driveway.

After college, Janice moved to Washington D.C. and was hired by the Culvers to help with childcare during a campaign year. Brad and I were gone a lot of the time, but in the summers, Janice brought the Culver kids to the farm, sometimes for weeks at a time. The Culvers approved. Living in Washington D.C. most of the year, they wanted their children to have some experiences in rural Iowa.

This was when Dad really took on Chet, initiating him into farm life, even buying him a mini-bike to ride around the farm—

an unfairness Brad and I complained about, as he'd never bought one for us when we were kids. "You had horses and ponies," Dad would say, "and we put up with all those wild animals of yours, quit complaining."

When I think about it, those summers at the farm hanging out with Del and Catherine Wilson were probably a good prerequisite to becoming the governor of Iowa. What better place to learn about the historical struggles farmers went through and their love of the land. Chet must have heard how we'd struggled through wet years and droughts, good markets and poor, and depressions of the past. Perhaps he got a feeling for how dramatically farming had changed. To survive in the 1980s and beyond, farmers had to buy more land, and acquire more expensive machinery, while still enduring low grain prices. Chemicals showed up to control weeds and pests and fertilize the soil for record crop yields, but some chemicals inadvertently contaminated drinking water wells. Topsoil was also lost with these new farming practices, and animals were taken off the land in favor of rearing them in huge confinement buildings that housed thousands of hogs, cattle, and chickens. These practices resulted in manure disposal problems and health concerns around hormone and antibiotic laden meat products.

My brother and sister's careers were deeply rooted in farming issues. Brad was a farm activist, using his folksy songs and poetry to communicate complex farm policy. He is a brilliant analyst of the Farm Bill and farm politics. As the eldest, Janice was the first of us kids who could reach the foot pedals on the tractor, but she spent even more time in our huge vegetable garden, helping mom plant, weed, and harvest. Chet would have worked alongside them, picking beans or digging potatoes. Janice's career has focused on organic and local foods, helping organic farm movements to create markets for small farmers. In many ways what Chet observed at our kitchen table, was that to farm successfully you have to respect the land. I'm pretty sure that's why the governor of Iowa drove all the way across the state that day to attend a corn farmer's funeral.

Epilogue - Keep the Fire Burning:
A Visit with Leopold's Ghost

In the early 1980s, when I was stationed at Mercer, an elderly gentleman with a cane came into the ranger station and asked our secretary if he could talk with the "wildlife guys." I think his name was Frank. He told me he was moving to a retirement home and was sorting out his things, many of which were going to the Mercer town dump. He had found a book and wondered if we wanted it. He told us he had worked in a fire tower as a young man, spotting forest fires for the Conservation Department, a job that is now done mostly with aircraft, which eliminated the need to staff the towers.

He handed me the book, saying, as it didn't teach much about hunting, fishing, or trapping, he hadn't gotten much out of it. It was a little too flowery for him. He'd heard the guy who wrote it was now famous and that it was too important to throw in the dump, did we want it? I looked down at the book and smiled. It was *A Sand County Almanac*, by Aldo Leopold. "Yes," I said, "I've read this book. It was assigned reading for a course I took in college. I've already got a copy, but I'll pass it on to one of our student interns who would be happy to have it."

I didn't try to explain to Frank what the book was about. If I had, I would have told him that, no, it wasn't about hunting or fishing or trapping, though Leopold had hunted, fished, and trapped most of his life. It was about a "land ethic," about how ecosystems function and the responsibility people have to live in balance with nature. I thanked Frank, and as we walked to his car, I said, "By the way, did you know that Aldo Leopold died fighting a forest fire?"

Flash forward to 2016. I was up early making coffee at our cabin on the Turtle Flambeau. Terry was still sleeping, but I was up because I was mad as hell.

The night before, I'd read an article (one of many that year) about legislation gutting Wisconsin's conservation laws. Over the past 10 years, I had seen draconian changes in environmental laws which favored a short-term economic approach to managing natural resources. Many were changes only obvious to policy wonks, being buried in the weeds of government process, and for the most part never received much press or public outcry.

One example was the elimination of the forestry mill tax, which cost about $24 per Wisconsin household each year and supported our DNR forestry program. In the northwoods, forestry is one of the driving forces of our economy, but this law phased out the segregated funding for DNR forestry programs. This meant that in the future, forestry work would be in competition with other taxpayer priorities like schools and roads. We also saw cuts to the state's Stewardship Program, which paid for the acquisition and maintenance of public lands. Local control of zoning decisions like shoreland preservation (good for water quality, loons, tourists, fishing) were curtailed. Many scientist positions were cut, including my former boss Dr. Mike Meyer, whose research had brought in grant funding to the state for many years. Top DNR leadership was hired with little or no regard for natural resources expertise. DNR staff had been "gagged"—prevented from speaking to the press or public without official sanction or from speaking on legislative proposals without express invitation. In my opinion, and the minds of many colleagues, the department had a continued focus on deregulation, cuts to environmental enforcement, and when economically convenient, a blatant disregard for science.

Education was in the crosshairs, and our DNR website managers had been directed to scrub all mention of climate change. Imagine that: foresters, who plan for forest harvests 50–100 years into the future, would not be allowed to consider the fact that some tree species such as red pine, paper birch, and aspen would not grow well in northern Wisconsin under climate change projections.

Closer to home, deer management was really taking a hit, with leadership hiring a deer "expert" from Texas whose recommendations included scrapping our scientifically-based deer registration system which prevented collecting deer age data. This was an important component of our population model that was used to set harvest quotas. These changes in baseline data collection seemed counter-productive, particularly in light of the emergence of chronic wasting disease, a fatal prion illness that had been discovered in southern Wisconsin whitetails and was spreading. This was a time when we needed accurate population information more than ever.

I'd even heard one northern Wisconsin state senator claim the reason he had led efforts to eliminate DNR science positions was because the DNR had destroyed the deer herd in the northwoods. This really burned me up because DNR data had long ago established the fact that white-tailed deer in the north are largely controlled by winter severity. Of the 10 most extreme winters ever recorded (deep snow and extreme cold), five of them had occurred within the last 10 years, with the winter of 2013–2014 being the worst. I'd aged thousands of harvested deer in my 23-year career in wildlife, so I knew for a fact that very few deer lived beyond six years of age. The deer population decline had little to do with over-harvest or high harvest quotas. As the coffee brewed, I muttered, "Why didn't the public know how terrible these changes would be? Wasn't the hunting public outraged?"

By the time Terry woke I was on my second cup of coffee, and the caffeine hadn't done much to soothe my mood. I told her I was going to call our friend Randy Jurewicz, who lived in Madison and closely followed the politics of natural resources. I wanted to share my sense of outrage and see if he thought there was anything to be done. When we spoke, what Randy said was, "Calm down, there's nothing we can do with our current leadership." I countered that the legislators must not fully understand the implications of what they were voting for. When my phone conversation with Randy ended, we agreed on one thing, Aldo Leopold must be turning over in his grave.

A few days later, the Trump administration announced the appointment of Scott Pruitt, a well-known climate change denier and fossil fuel advocate, to be head of the U.S. Environmental Protection Agency. Terry and I had another gloomy coffee hour, considering the weakening of state and now federal protections for the environment. In the past few years when a law gutting environmental regulations in Wisconsin passed, citizens would say, "Don't worry, the federal government will protect the water, air, and wildlife." While not entirely true, there was a small comfort in knowing that at the national level, conservation was still on someone's radar. Now, it seemed that the Wisconsin model of deregulation and cuts to science budgets were being implemented at the federal level.

Terry headed off to swim laps, an activity she finds particularly meditative. She came back from her swim full of energy and said that she had an idea. She suggested that instead of feeling glum and defeated, why not contact our friends and colleagues, many of whom were recently retired natural resource professionals, and form a group that would be able to share science with the public without constraints? The group, she said, would be non-political. We would simply use our expertise to comment on issues and policy proposals as they came up. Our goal would be to serve citizens, media, and legislators who wanted unbiased scientific information on conservation issues. Terry said, "I've even got the perfect name for our group–Green Fire."

Green Fire refers to a much-quoted section of an essay in *A Sand County Almanac*, called "Thinking Like a Mountain." In the essay, Leopold writes about a job he held as a young professional while working for the federal government in the mountains of New Mexico. His role was to assist ranchers grazing cattle on public land by eliminating predators. He describes having a life-changing experience after he shoots a wolf and watches the "green fire" fade from her eyes as she dies. At that moment Leopold has an epiphany, realizing that predators, deer, and habitat are inter-connected. It was a pivotal moment in the development of his "land ethic." Thinking like a mountain meant understanding the role and value each part of an ecosystem plays. The green fire symbolizes this epiphany. I immediately agreed: Green Fire, that's it!

Within weeks, Wisconsin's Green Fire was born. Our first meeting brought together around 40 mostly retired scientists and natural resources experts from the DNR, the University of Wisconsin, the Environmental Protection Agency (EPA), and U.S. Forest Service to name a few. By the following fall, we had organized and planned our first annual meeting, which included a field trip to Aldo Leopold's famous "shack" in Wisconsin's Driftless Area. The shack was Leopold's family cabin. It is the setting for much of *A Sand County Almanac*, and many of his essays are based on the story of this humble structure located on a worn-out farm where Leopold and his family put his land ethic into practice. Today, the shack is surrounded by massive oaks and pine that the Leopold family planted decades ago. On a beautiful Friday in September, the day before our official membership meeting to elect officers and a board of directors, we toured the Aldo Leopold Center grounds with Curt Meine, a renowned Leopold scholar, and then headed to the shack for a visit.

To me it was something like visiting a shrine, and as this was my first visit to the shack, I was hoping for a contemplative moment on my own. I purposely waited until the group before me had entered and exited before I stepped quietly into the tiny building. The shack was originally a chicken coop, and its homely origins were still visible. I studied the design, noting how Aldo had added on bunks for sleeping. There was a rustic fireplace and a rough-hewn table where I could picture him sitting and pondering the essays for his almanac.

I am generally not too sentimental, but there in the shack by myself I felt strangely moved, almost as if Leopold was close at hand. I'm not much of a believer in ghosts—my background is in science. In science we test hypotheses, collect data, provide analysis and conclusions, then send our work out for peer review. It's hard to do that with a spirit. But I thought that if Leopold were to revisit Terra Firma it would be at a place he knew and loved. After all, he died just a few miles from here fighting a neighbor's brush fire.

So, a bit shyly I said, "Okay, Aldo, if you're here, we could use some help. Your legacy is in peril." I sat in silence waiting. No ghost appeared. No table moved...no rattling sounds emanated from the floor or walls...only silence. After a few minutes I got up from the table, walked out the door of the shack and closed it behind me.

The next day over 90 of our state's active and retired natural resource professionals came from all corners of the state. Among them were retired game wardens, fish managers, educators, water regulatory folks, former DNR administrators, division chiefs, journalists, and a former DNR secretary. The daughter of former Senator Gaylord Nelson (world renowned as founder of Earth Day), Tia Nelson, was our keynote speaker. People stepped forward to run for the board, to lead committees, and to make a stand for the conservation legacy of Wisconsin. There was a profoundly powerful spirit within that room—a spirit to continue our public service to Wisconsin. As we endorsed the name, Wisconsin's Green Fire – Voices for Conservation, I thought that Leopold's "land ethic" was there in the room, empowering our collective hopes for a better future.

In the first year, our membership grew to over 200 and at this writing in 2022, we're at nearly 850 supporters. Terry is the board president and we have hired staff. We have 11 technical work groups and can frequently be found at the state capital providing scientific testimony on proposed bills. We have stuck to our mission, to provide sound science to decision makers and the public and I believe our voices are making a difference.

When I look back over my four decades of involvement in resource management, I can see that we were trying to follow the teachings of Leopold and other conservation pioneers. Our focus was to put damaged ecosystems back together and build a love for the land in future generations. We started our careers in what was a landscape impacted by the cutover of original forests and subsequent wildfires. Wisconsin had lost around half of its wetlands and massive amounts of soil from poor farming practices and the dust bowl years. We had seen impaired waters and air from toxic chemicals and pollution and loss of species that occurred before passage of the Clean Air Act, Clean Water Act, and Endangered Species Act.

At the time, these problems seemed daunting if not impossible to solve but we worked hard to make headway. Now we seem to be facing even more impossible challenges, many global in scale. Our aquifers are diminishing, drinking water in our state continues to be impaired, forested land is becoming more and more fragmented as land ownership patterns shift, shoreline development has moved from small camps and cabins to large second homes, and chronic wasting disease is spreading in our state's white-tailed deer herd. There are also a few elephants in the room, climate change, racial justice, and ecological sustainability. In the face of need for a massive shift to clean energy and a reduced carbon footprint, many of our leaders are choosing to turn a blind eye, ignoring science at the peril of future generations. Inequity seems to be growing and short-term economics still holds the trump cards.

I wrote this book to share stories of my career, to revisit an era when we worked closely on the land. Perhaps most importantly, I wrote this book to encourage future conservation leaders to take up where we have left off. Leopold's land ethic has never been more important. In an age of increased separation from nature we need to find new ways to help humans and future generations connect to and value the beauty and complexity of this planet.

The year 2020 saw thousands of school children in Great Britain and Australia skip school to protest government inaction on climate change. Their pleas for adults to act now were impassioned and moving. My then 14-year-old granddaughter, Jalina, won the local conservation speech contest at her school with a talk on climate change, but she was shot down at the regional level when a judge said that her speech was based on opinion, not science. That adult may have hoped to quell her youthful dedication to fighting climate change, but in fact it added to her motivation to make leaders accountable for the future.

When I visited Leopold's shack I sought inspiration. The next day I thought that perhaps Leopold's ghost was with us as we searched for direction for our new group. Every generation has inspirational leaders, and for me, visionaries like Aldo Leopold, Sigurd Olson, Fran Hamerstrom, Ernie Swift, and Gaylord Nelson filled that role. Beyond those who achieve fame, every human knows people who have been mentors, quietly leading by example in small daily doses. I have been fortunate to have known many such individuals as friends and colleagues.

In this book, I wanted to share some of the wisdom I gleaned from my mentors as well as a sense of their love of the land. Memories of the great times we spent together serve as a well of inspiration for me which I can dip into when I need to be rejuvenated. My moment at Leopold's shack was a bit like going to church; I wanted to gain a feeling of communing with a greater wisdom. However, the shack remained quiet, with the dust motes sparkling in the late afternoon sun. I think Aldo spoke to me the next day, when the gathered energies of my many friends and mentors came together to continue our work for the land. Maybe I do believe in ghosts, not as eerie spirits from beyond, but as inspired messengers, encouraging us to keep up the good fight for the land, for wildlife and wild places, and for a sustainable human existence for future generations on this beautiful Earth.

This is all sounding pretty darn serious, so I would like to end this lengthy tale with a last word of advice from wilderness advocate and overall rabble-rouser, Edward Abbey, advice that speaks to my heart and that I suggest others to follow as I have.

"One final paragraph of advice: do not burn yourselves out. Be as I am—a reluctant enthusiast . . . a part-time crusader, a half-hearted fanatic. Save the other half of yourselves and your lives for pleasure and adventure. It is not enough to fight for the land; it is even more important to enjoy it . . . so get out there and hunt and fish and mess around with your friends . . . bag the peaks, run the rivers and contemplate the precious stillness. . . . I promise you this one sweet victory over our enemies, You will outlive the bastards."

Acknowledgments

Like my cabin, this book was really a gift from my family and friends. They made it all possible, whether starring in a story or working behind the scenes as the book evolved. My dear wife Terry gets a special thanks for sticking with me through all the edits and for creating illustrations that helped dramatize our story. Deep appreciation is also due my family who listened to me revise these tales over the years, especially my son Zach, daughter in law Melody, and grandkids Lola, Keewin, and Jalina, who when very young would beg, "Tell us another story Grandpa." Once they hit teenage years, their requests for a tale grew fewer, but even then, they were still susceptible to my ramblings, in other words they learned to tolerate another story. My sister-in-law Diane was a top-notch editor, critic and patiently led our promotion efforts. Without Ellen Akins, my first editor, best critic and fan, we would never have had the courage to launch this effort. Cornerstone Press editor Jeff Snowbarger, director and publisher Dr. Ross Tangedal, and the entire crew at Cornerstone, and designer Jerry Lehman, polished the book and brought out the best in these stories! My tales were fact checked by many science colleagues including: Bruce Bacon, Mike Meyer, Kevin Kenow, Sumner Matteson, Randy Jurewicz, Fred Strand, Adrian Wydeven, Dick Theil, John Olson, Pat Manthey, John DePue, Cheryl Dykstra, Charlie Zinsmaster, Christa Reinert, Lisa Hartman, Dan Haskell, and Jerry Belant. Hannah Gargrave and Jeff Richter were test readers for the manuscript, and friend and fellow writers John Bates and Tom Joseph weighed in with invaluable editorial suggestions. My brother Brad and sister Janice made sure my childhood recollections stayed on track.

I have had the good fortune of learning from dozens of teachers; men and women older and younger than me, who taught me the art and methods of wildlife conservation. Some of these gifted people appear in these stories but many more helped me learn along the way. This list of mentors, friends, and colleagues is far too long to catalog here and if I did, I fear I might miss someone.

And lastly, I want to thank my mother and father who taught me to love nature. Looking back over my career, I think the most important lesson in the end was to pass on what I learned to others, and in particular to young people, with joy and dedication, for they are the hope of the future.

Partial Cast of "Characters"

(in order of appearance)

I thought that, rather like a Russian novel, where the reader is likely to get turned around in lists of names, a bit of assistance in the form of a cast of characters might be of service. I wondered too, if it might also quench the curiosity of those who wonder where all these colorful individuals ended up. I definitely owe all these folks a resounding "thank you," for without their wisdom, skill, and dedication, my life would have been far less rich, exciting, and full of adventure.

Sam Moore – after retirement from the DNR in Hayward, Sam was a regular guest discussing wildlife issues on the public radio station, WOJB.

Jack Koch – failing at his first attempt at retirement he returned to work another 10 years as a seasonal before retiring a second time. He can still be found from time to time at the DNR office in Park Falls around coffee break time.

John Olson – retired as State Furbearer Biologist, after representing Wisconsin in such far-flung locales as Siberia. He continues to travel the country and world teaching trapper education (Trapping Matters).

Jim Cox – retired from DNR fish management. You can find Jim trolling for lake trout out of Saxon Harbor on Lake Superior.

Anna from "Dancing in the Moonlight" – went into the restaurant business and makes the best pasta in the northwoods.

Red Fox from "The Fox Whisperer" – who knows, I still wonder about that fox.

Northwoods Nessie – occasionally seen poking its head above water on Long Lake.

Fred Strand – retired as a wildlife supervisor and has commenced a new career as a Bayfield County Board member. He still helps out with common tern management.

Sumner Matteson – is an endangered species ecologist and has published a history of naturalists in Wisconsin called *Afield*, from Little Creek Press.

Zach Wilson – is a conservation specialist with Iron County's Land and Water Conservation Department and is still tracking American marten with area high school students.

Dangerous Dan Haskell – is partner to Amber Roth, became an expert on shoreland habitat restoration and currently lives in Bangor, Maine.

Lassoed Bear – went on to a charmed life deep in the woods, avoiding all contact with humans.

Dr. Mike Meyer – retired from the research bureau of DNR, chairs Wisconsin's Green Fire's Science Council, and owns a biological consulting business called NOVA Ecological Services.

Bruce Bacon – is a retired DNR wildlife biologist, and volunteers on various birding projects including banding eagles and songbirds.

Randy Jurewicz – is retired from the DNR endangered resources and serves as a council member for the Timber Wolf Alliance.

Christa Reinhart – is the owner/manager of Loony Beans Coffee and Bait in downtown Mercer. It's the only place in Mercer you can order a cappuccino and a cup of leeches at the same counter. You don't want to get them mixed up.

Rib Breaker – passed away in 1996 but watch out for his offspring if you are out and about on Wisconsin lakes.

Lisa Hartman – lives on a farm near Portage, Wisconsin with her partner Mike Mossman. She plays folk music, teaches English as a second language, and gives educations presentations with her educational turkey vulture, Babette.

Adrian Wydeven – is retired from DNR, co-chairs the Wisconsin's Green Fire wildlife work group, and volunteers with the Timber Wolf Alliance.

Terry Daulton – has served two terms as president of Wisconsin's Green Fire. She paints landscapes, coordinates an arts and science program at UW–Trout Lake, and is the illustrator of a very fine book or two.

Colleen Matula – is recently retired after serving many years as Wisconsin DNR's northern region forest ecologist.

Jack Massopust – passed away in 1985. A scholarship at UW-Stevens Point was established in his honor.

Bruce Kohn – retired from DNR Bureau of Research, still assists with bear denning research at UWSP.

Charlie Zinsmaster – after retiring from DNR, he served as Iron County Forest Administrator for several years.

Girl Scout Troop 237 – never returned to the Apostle Islands. Future outings were directed elsewhere.

Bandit the dog – though nominated, did not receive "Dog Hero of the Year" award.

Dave "Fud" Evens – is still running a raptor banding station on Minnesota's north shore.

Dr. Paul Strong – recently retired as the forest supervisor for Chequamegon/Nicolet National Forest in Northern Wisconsin.

Dr. Jerry Belant – has taught at several universities and currently The Boone and Crockett Chair in the Wildlife Conservation Department at Michigan State University. He can sometimes be found collaring wolves on Isle Royale National Park or lions in Kenya.

Dave Evers – runs a biological research firm called Biodiversity out of Maine. He is still capturing and banding loons.

Monica Joseph (who peeled the cabin ridge pole) – is a wildlife biologist stationed in Crystal Falls, Michigan.

Loon chick from "Close, but No Cigar" – was recaptured as an adult five years later, one mile from the original nest territory and was still wearing its web band.

Alan "Ski" Sabatke – went on to build a dozen more Scandinavian-scribed log homes. Currently he has a custom door business called Dogwood Doors.

Brad Wilson – is a retired farmer who is still promoting small family farm values and sustainable agriculture. He has a foundation called "Fireweed."

Wild Bill Seybold – teaches young people hunting and fishing recreation for the Idaho Fish and Game Department. He is also a game warden.

Dr. Cheryl Dykstra – is the editor of the journal *Raptor Research* and is a wildlife consultant doing a long-term study of red-shouldered hawks. Her team has banded over 2,500 nestlings and 300 adults.

Keith Warnke – went on to become a Division Administrator at the DNR Central Office in Madison before retiring in 2022

Kevin Kenow – is an emeritus researcher at USGS and continues to work part-time on selected projects.

Steve Thielking – when I last heard he was studying fox at the Channel Islands National Seashore in California.

Ken Pemble – is a wildlife biologist for the Chequamegon Nicolet National Forest.

Dr. Claire Gower – received her U.S. citizenship in 2019. She is currently a biologist for the Montana Department of Fish and Game.

Diane Daulton – retired from her natural resource career but continues to work as a part-time naturalist at Copper Falls State Park. She authors "The Water Column" (*The Bottom Line News*) and enjoys equine and canine pursuits.

John Depue – is a wildlife biologist with Michigan DNR stationed in Hancock.

Dr. Amber Roth – is a professor at the University of Maine-Bangor with a specialty in golden-winged warblers.

Ruth Bennett – has finished her doctoral dissertation at Cornell University Lab of Ornithology.

www.ingramcontent.com/pod-product-compliance
Lightning Source LLC
Chambersburg PA
CBHW022045020426
42335CB00012B/555